The VHF GMDSS Handbook

Free update service

The GMDSS is new, and is sure to undergo a good deal of 'fine tuning' as it beds down. If you would like to receive a free update, please send a Stamped Addressed Envelope to:
Fernhurst Books, Duke's Path, High Street, Arundel, BN18 9AJ

The VHF GMDSS Handbook

A Manual for
the RYA 'VHF only' Certificate,
the CEPT Short Range Certificate and
the GMDSS RESTRICTED OPERATOR'S Certificate

J. Michael Gale

© Fernhurst Books 1998

Fernhurst Books,
Duke's Path, High Street, Arundel,
West Sussex, BN18 9AJ, UK
Tel: 01903 882277
Fax: 01903 882715

Printed and bound in Great Britain

The moral right of J. Michael Gale to be identified as the author of this work has been asserted in accordance with ss 77 and 78 of the *Copyright, Designs and Patents Act* 1988.

British Library Cataloguing in Publication Data:
A catalogue record for this book is available from the British Library

ISBN **1 898660 54 9**

Acknowledgements:
The author and publishers would like to thank Mike Peyton for drawing the cartoons and ICOM (UK) Ltd (sole importers of ICOM radio communications equipment) for their assistance in the preparation of the book. Thanks are also due to Kim Fisher of the Marine Safety Agency and to Kelvin Hughes Ltd for their advice on the manuscript.

Photo credits:
ACR Electronics Inc: pages 56 and 57; Coastguard Headquarters: pages 2, 11 and 23; Fernhurst Books: pages 14, 18 and 39; Michael Gale: pages 19 and 20; ICOM: page 27; ICS Electronics Ltd: front cover page 52 and page 55; Marine Safety Agency: page 49; McMurdo Marine: pages 50, 58 and 59; Navico Ltd: page 60 and the back cover

Design & typesetting by John Carden
Cover design by Simon Balley
Printed and bound by Hillman Printers, Frome, Somerset, UK

Michael Gale is also the author of *Marine SSB Operation* (GMDSS edition) (ISBN 1–898660–40–9) and *Marine VHF Operation* (GMDSS edition) (ISBN 1–898660–39–5)
For a free, full colour brochure on these, our book entitled GMDSS *for Small Craft* (ISBN 1–898660–38–7) by Alan Clemmetsen, and all our other nautical titles, please write, phone or fax the publishers: Fernhurst Books, Duke's Path, High Street, Arundel, England, BN18 9AJ Tel: 01903 882277 Fax: 01903 882715

The photographs facing the Title page are of the Falmouth Maritime Rescue Co-ordination Centre

All rights reserved. No part of this publication may be reproduced, stored in a retrieval system, or transmitted in any form or by any means, electronic, mechanical, photocopying, recording or otherwise, without the prior written permission of the publisher.

Set in 9/11pt Novarese

CONTENTS

Foreword 6
1 Certificates and Licences 7
2 The Phonetic Alphabet 10
3 Organisation of the Marine VHF Band 12
4 Who can you talk to? 18
5 Some Basic Technicalities 24
6 Priority of Signals on Channel 16 28
7 Ship-to-Ship Communication 29
8 To Call Coastguards 32
9 To Call Yacht Clubs, Marinas and Port Radio Stations 37
10 Ship-to-Shore Telephone Calls 38
11 Distress Signal (MAYDAY) 44
12 Acknowledgement of Distress Signals 46
13 The GMDSS – Introduction 47
14 The DSC Controller 51
15 NAVTEX, EPIRBs and SARTs 54
Important Procedural Words (PROWORDS) 61
Glossary 62
Useful addresses 63
Bibliography 63

FOREWORD

We live in an era of great and rapid change. As the end of the 20th Century approaches, it seems that every facet of human activity is being changed and none more so than Marine Radio. Such is the pace of change these days that any textbook, or even a monthly technical journal, is out-of-date as soon as it is published. In the 10 years since my first book *Marine* VHF *Operation* was published in 1987, it was revised twice with over 100 updates each time! For example: with the recent proliferation of cheap mobile telephones which can be used in inshore waters, it is likely that VHF Coast Radio Stations could be closed-down in 1999, as little use is being made of them.

Marine Radio has kept pace with advancements in electronics technology over the last 100 years. However, the next step is a big one called the **Global Maritime Distress and Safety System** (GMDSS). It is a completely new system of **calling** and **Distress Alerting** using new technology called **Digital Selective Calling (DSC)**. As well as being a more efficient and reliable system, it obviates the necessity of keeping a listening watch. It is the radio equivalent of the change from sail to steam – but over a much shorter period! However, it must be emphasized that **traditional R/T message procedures and channels** are not being changed.

As the GMDSS does not become fully operational until 1st February, 1999, we are presently (1998) approaching the end of a transitional period between two very different systems. Although small craft under 300 grt are not obliged to 'join' the GMDSS, it will be prudent to do so in the interests of safety. To ease the transition, this book is in two parts. Chapters 1–12 are similar to the same Chapters in my book *Marine* VHF *Operation* (the 1997 edition) and cover traditional procedures. Chapters 13 to 15 cover the GMDSS. These will become relevant from late-1998 when the new-type GMDSS VHF sets (Class D) become available.

Like most things which seemed like a good idea at the time, the GMDSS is sure to undergo 'fine tuning' in the 21st Century after it has had time to settle down with experience. In the meantime, this book covers the transitional situation regarding small-craft Marine VHF radio and covers the theoretical syllabus of the professional GMDSS *Restricted Operator's Certificate* (ROC) and the leisure craft 'VHF *only*' *Certificate* and the *Short Range Certificate* (SRC) exams. To this end, throughout the book, a '(Q)' follows every statement which answers an ROC, VHF or SRC examination question. Consequently, particular attention should be paid to these statements.

Finally, may I express my very considerable thanks to Kim Fisher of the Maritime and Coastguard Agency for the time and trouble he took reading the manuscript of this book.

J. Michael Gale
The Radio School
Hayling Island

1

CERTIFICATES AND LICENCES

Like motor vehicles, two documents are required to install and operate marine radiotelephones.

1 SHIP'S RADIO LICENCE

The Ship's Radio Licence equates to the motor vehicle licence or tax disc. Although the UK licensing authority is the Radiocommunications Agency (Q) (an Executive Agency of the Department of Trade and Industry), Ship's Radio Licence applications are obtained from and returned to Wray Castle Ltd., in Ambleside – their address will be found in the List of Useful Addresses at the end of the book. Ship's Radio Licences are valid for one year from the date of issue but, unlike the car tax, are not transferrable (Q). When a boat is sold, any Radio Licence immediately becomes invalid and there is no refund for the time remaining. The new owner must apply for a fresh Licence and the vendor must apply for a fresh Licence for any replacement boat. The annual fee is currently (1998) £22. A reminder, with a remittance slip, is sent about four weeks before the expiry date.

Call-sign

In just the same way that motor vehicles are allocated a Registration Number when first licensed for the road, a similar system applies to radio stations. Termed a **call-sign**, it identifies the station uniquely by voice or Morse code (Q). Once allocated to a vessel, it stays with that vessel so long as it remains under the original flag. It is not transferable and is only withdrawn if the vessel is sold into foreign hands.

Radio call-signs incorporate two simple codes:
a The first digit (or two) indicates the nationality of the station, e.g., all British call-signs start with G, or M, or 2.
b The number of succeeding letters indicate the type of radio station concerned. British marine shore stations have two further letters; for example, Niton Radio's call-sign is GNI. British ships have three further letters (sometimes followed by a numeral) e.g., MXYZ6, whereas British aircraft have four further letters following a G–, e.g., G–BAYP.

Unlike motor vehicles and aircraft, marine call-signs do not have to be displayed although there is no reason why they shouldn't be. Fishing boats often paint their call-sign on the wheelhouse roof where it can be seen by aircraft. The Author put 2JPG on the mainsail of his ketch in letters 2 feet high and in letters 18 inches high on the mizzen.

Although vessels **can** be identified by their name (Q), use of the call-sign is preferable when communicating with a stranger (most shore stations). It is absolutely unique; it identifies the vessel's flag and is easily copied by anyone of any nationality **because it is always spoken (in full) phonetically**. The Phonetic Alphabet is covered in detail in Chapter 2.

SHIP MMSI

For DSC purposes, a third method of identifying **ships** and **marine shore stations** (only) has recently been introduced. Quite simply, it is a nine digit 'telephone number' called a **M**aritime **M**obile **S**ervice **I**dentity or **MMSI** (Q) which is automatically included with every DSC call. It contains three codes. The first three figures (called Maritime Identification Digits) are the country code. The United Kingdom has been allocated three country codes on account of the size of the register which includes leisure craft. They are 232, 233 and 234 (Q), e.g., 232654000. One, two or three 'trailing zeros' indicates a facility for local, national or world-wide automatic public telephone calls respectively (Q).

The MMSI is allocated (free) by the National Radio Regulatory Authority (Q) when application for a Ship's Radio Licence is first made or at its renewal (Q). It is programmed into the new VHF

The VHF GMDSS HANDBOOK

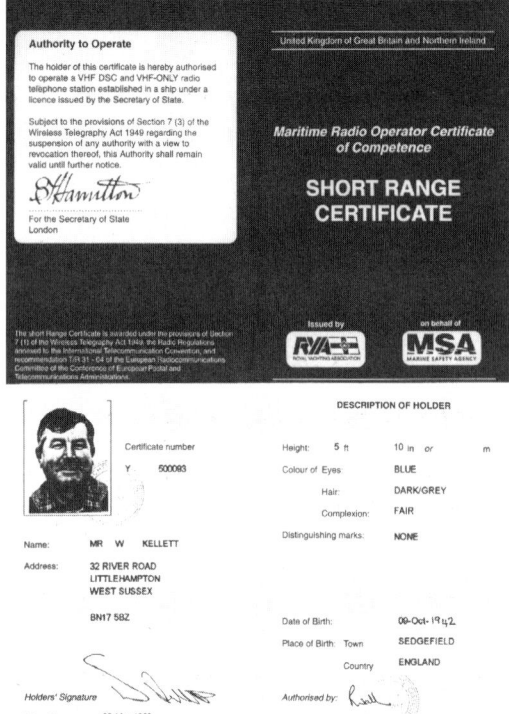

The operator's Short Range Certificate (left) and the boat's Ship Radio Licence (right)

sets (or DSC controllers) by the manufacturer (Q) and cannot be altered by the user. Thus, the new VHF radios with DSC facilities may only be obtained by licence-holders with an MMSI. As this makes the new marine radios unique to a particular vessel, they must be reprogrammed by the manufacturer if obtained second-hand.

Group MMSI

In addition to an individual MMSI, ships may **self-programme** a 'group' MMSI if belonging (even temporarily) to a particular group or fleet, e.g., a race, rally, cruise-in-company or fishing fleet. Group MMSIs are also allocated by the Radio Regulatory Authority (Q) and are distinguished by a 'leading zero' – for example **0**23200583 (Q). This enables a special broadcast to be made to the whole fleet.

Shore Station MMSI

These are distinguished by two 'leading zeros', so, for example the MMSI of Solent Coastguard is **00**2320011 (Q).

TRANSPORTABLE SETS

These may be covered by the parent Ship's Licence if used *only on board one particular vessel* or *in that vessel's tender or lifeboat* (Q). In such a case, both stations use the vessel's name with the master station adding the suffix 'Control' and the sub-station(s) adding the suffixes Alpha, Bravo, Charlie, etc., as allocated by the skipper. For example, *Pandora Control, this is Pandora Alpha* (Q). It should be noted that Channels 15 and 17 are reserved for this type of communication as the power output of the VHF set is then automatically reduced from the maximum of 25 watts (Q) to 1 watt. This reduces the possibility of causing unnecessary interference to other users operating on the same channel (Q).

If the transportable is needed for use on different unlicensed vessels, a separate Transportable Licence is required (Q). In this case, transportables are identified by a 'T' registration number (or special MMSI), not a call-sign. Although British Coast Radio Stations will accept transportable registration numbers for public telephone calls using the 'YTD'

The VHF GMDSS HANDBOOK

system of payment, foreign Coast Radio Stations will not! Transportables may have limited facilities for DSC Distress Alerting (see page 50).

Note: The use of marine radios ashore is strictly forbidden without an appropriate Licence. (Training establishments must register with the National Radio Regulatory Authority for permission to demonstrate transmitters into a 'dummy load' or artifical aerial.)

2 OPERATOR'S CERTIFICATES

In addition to the Ship's Radio licence, marine radios may only be **controlled** by the holder of a *Certificate of Competence and Authority to Operate* (Q). However, anyone on board **may use** the radio under the supervision of a qualified operator (Q).

'VHF only' Certificate

From 1903, the GPO (and then BT) set all the various grades of radio operator exams and awarded the Certificates on behalf of the National Radio Regulatory Authority. For several decades, the marine VHF qualification for 'voluntarily-equipped' vessels (i.e., yachts) has been the 'Restricted Certificate of Competence in Radiotelephony (VHF only)'. The RYA have held the contract for the exam since April, 1983. It consists of a 30-minute, 15-question written paper and a short practical test on a simulator. At the time of writing (May 1998), this qualification can still be obtained. However, it will have limited application from 1999 and will soon be withdrawn and replaced by the Short Range Certificate.

CEPT Short Range Certificate

In preparation for the declaration of an A1 Sea Area off Britain's South Coast by HM Coastguard on 1st January, 1998, the RYA introduced the new CEPT (*Common European*) *Short Range Certificate* (SRC) for voluntarily-equipped craft using VHF on 1st September, 1997. For the old 'VHF only' Certificate, the RYA's contract was for the exam only; the then National Radio Regulatory Authority did not make attendance on a course mandatory. Now that the Maritime and Coastguard Agency (MCA) are the certifying authority for British marine radio operators, attendance on an RYA-recognized (usually weekend) course by an RYA-qualified Instructor at an RYA-recognized Centre is mandatory for the new SRC. Even the Author, who held VHF courses for 10 years before the RYA became interested, had to attend a one-day RYA Instructor's course to be authorized to teach the SRC!

The full exam consists of a written paper, an R/T exercise, and a practical test on a Class D DSC controller or VHF/DSC computer simulator. The paper, for which 40 minutes is allowed, contains 25 questions. The first six require short answers but the remainder are multiple choice. The pass mark is 75% for the first six Distress and Urgency questions in Part A, and 60% for the remaining questions in Part B. There is no time limit for the R/T exercise or the practical DSC test.

SRC Upgrade Course and Exam

Holders of any of the old non-GMDSS Certificates can upgrade to the new SRC by attending only the second day of the course – usually held on a Sunday – at a reduced course and exam fee. They tackle only the first 15 questions on the written paper (30 minutes) and the practical DSC test for which there is no time limit.

Restricted Operator's Certificate (ROC)

For professional deck officers sailing on ships that are compulsorily equipped under SOLAS and sail within a Sea Area A1 *only* (such as the Cross Channel ferries), the minimum requirement is the higher-grade GMDSS *Restricted Operator's Certificate* (ROC). It was introduced into the UK by the MCA on 1st January, 1998. The Examining Body is the Association of Marine Electronics and Radio Colleges (AMERC) – please see the *List of Useful Addresses* at the end of this book. The syllabus is very similar to that of the SRC but to a higher standard. This book also covers the ROC theoretical syllabus.

Secrecy

Before the award of a Radiotelephone Operator's Certificate, all candidates must sign a *Declaration of Secrecy* not to 'improperly divulge to any person the purport of any message' which they may acquire whilst acting as a marine radiotelephone operator. (Q).

Enforcement

Enforcement of the Radio Regulations in the UK is carried out by the Radiocommunications Agency. They operate through a network of regional offices throughout the UK. The maximum penalty of infringement is presently (1998) an unlimited fine or 2 years imprisonment *plus* the forfeiture of all the equipment!

2
THE PHONETIC ALPHABET

For accuracy in communication, it is absolutely vital that every radio or telephone operator is fluent in the phonetic alphabet. When talking over the radio or telephone, it can be almost impossible to distinguish between 'F' and 'S', 'N' and 'M', or 'B' and 'P'. To ensure accuracy, a name or uncommon word may be spelled out using a word for each letter.

This technique originated in the First World War with the 'ACK-ACK, BEER-BEER' system. Veterans may remember the example used to illustrate the necessity of such a system when a message reputedly started out in a trench as 'Send reinforcements, we are going to advance' but ended up at Battalion HQ as 'Send three and fourpence, we are going to a dance'! Certainly some years ago, a student telephoned the author to ask exactly where, in Salisbury, a class was being held. When asked why he thought it was being held in Salisbury, he replied 'Your wife told me the address was Wickham Road, Sarum' (Sarum is the old name for Salisbury). To which the reply was 'No she didn't – she said Wickham Road, *Fareham*. I SPELL "FOXTROT, ALPHA, ROMEO, ECHO, HOTEL, ALPHA, MIKE" '.

The example quoted uses the current system, introduced on 1st January 1955 for NATO forces which replaced the Second World War system. Originally used for military purposes only, it became so successful that it is now used for *all* marine and aeronautical radio throughout the world (Q). Note that when spelling out a word phonetically, it is essential to say 'I SPELL' before launching into phonetics; this is a signal for the recipient to put pencil to paper and take down the message.

Phonetic numerals

There is also a system of International phonetic numerals but it is rarely used as most communicators throughout the world say numerals in English. However, there is still some possibility of

Phonetic Alphabet

Letter	Word	Spoken as	Variations	Letter	Word	Spoken as	Variations
A	ALPHA	AL–fah		M	MIKE	MIKE	
B	BRAVO	BRA–voh		N	NOVEMBER	no–VEM–bah	
C	CHARLIE	CHAR–lee	French may say 'SHAR–lee'	O	OSCAR	OSS–kar	
				P	PAPA	pa–PAH	
D	DELTA	DELL–tah		Q	QUEBEC	key–BECK	
E	ECHO	ECK–oh		R	ROMEO	ROW–mee–oh	
F	FOXTROT	FOKS–trot		S	SIERRA	see–AIR–rah	
G	GOLF	GOLF		T	TANGO	TANG–go	
H	HOTEL	hoh–TELL	French may say 'oh–TELL'	U	UNIFORM	YOU–nee–form	
				V	VICTOR	VIK–tah	
I	INDIA	IN–dee–ah		W	WHISKY	WISS–key	Germans may say 'VISS–key'
J	JULIETT	JEW–lee–ETT	Germans may say 'YOU–lee–ett'	X	X–RAY	ECKS–ray	
K	KILO	KEE–loh		Y	YANKEE	YANG–key	
L	LIMA	LEE–mah		Z	ZULU	ZOO–loo	

The VHF GMDSS HANDBOOK

confusion, particularly with 'five' and 'nine', so it is usual to shorten 'five' to 'FIFE' and extend 'nine' into 'NINER' to help distinguish between them. Here is the complete list:

Numeral	Spoken as
1	WUN
2	TOO
3	TREE
4	FOW–er
5	FIFE
6	SIX
7	SEV–en
8	AIT
9	NINE–er
0	ZERO

Do not say 'OH' for '0' as it can sound like 'TWO'. Numbers should be spoken figure-by-figure except for whole thousands, and if a decimal point occurs in a number, it is given as 'point'. For example:

Numeral	Spoken as
32	TREE TOO
459	FOW–er, FIFE, NINE–er
200	TWO, ZERO, ZERO
8162	AIT, WUN, SIX, TWO
7000	SEV–en THOUSAND
2.5	TWO POINT FIFE

Below: Spelling out the ship's name and call-sign in phonetics not only ensures accuracy; it also gives the radio operator at the other end the opportunity to write down each letter in sequence

3

ORGANISATION OF THE MARINE VHF BAND

To use your VHF set to full advantage (and pass the exam) it is essential to have a thorough understanding of the organisation of the marine VHF band which is quite different from any other radio band.

Radio 'frequency' means 'cycles-per-second'. A frequency of one cycle per second is expressed as one Hertz (1 Hz), after the German physicist and radio pioneer H. R. Hertz. For convenience, the whole range of radio frequencies is divided into a number of groups or 'bands' according to frequency range. Marine VHF operation is carried out within the Very High Frequency (VHF) band which ranges from 30 megahertz (MHz) to 300 MHz, that is 30–300 million cycles per second (between ten metres and one metre wavelength). If you have a VHF band on your domestic radio (it may be marked 'FM' or, if of German make, 'UKW') you will see that the dial is marked from 88 to 108. These are the limits, in MHz, of the *broadcasting* part of the VHF band – around three metres in wavelength.

INTERNATIONAL CHANNELS

The limits of what is called the International Maritime Mobile (IMM) Band are 156 MHz to 162 MHz. So, although we are operating in the same VHF band as the BBC, we are operating in a different part of it, which is why we do not hear the BBC on our marine VHF R/T – or ships on our domestic radio.

Between the limits of 156–162 MHz, the IMM Band is further sub-divided into a number of small segments or 'channels'. Each channel is numbered for convenience, because we are dealing in hundreds of megahertz to two or three decimal places, which even professionals have difficulty in remembering. The use of numbered channels makes for very easy operation of the set; simply turn the knob (or push the button) for the appropriate channel number. The actual frequencies involved are of purely academic interest.

The channelling of the IMM Band was rearranged on 1st January 1972 which makes the present scheme look highly complex and very illogical. To make for easier understanding, consider first the original scheme which was introduced soon after the Second World War. As you see from Table 1, the channels were numbered consecutively from 1 to 28 (although there is a Channel 00 on 156.000 MHz, it is for HM Coastguards' own private use). In this respect, the IMM Band is like the present Citizen's Band (CB) which is similarly divided into 40 numbered channels. However, the big difference lies in the fact that whereas every CB operator is fully authorised to use *any* of his 40 channels for any purpose, the marine VHF operator has to choose a channel which is appropriate for the use intended. As you see from Table 1, every channel in the IMM Band is dedicated to a particular use. Some channels are dedicated exclusively but others are shared (Q).

In ship-to-shore communication, the shore station dictates the particular channel (Q) but with inter-ship communication, it is the *station called* which nominates the channel (Q). Thus *it is only with inter-ship operation that the on-ship operator ever gets the chance to nominate a channel.*

Channel 16

From Table 1, you can see that Channels 06, 08, 09, 10, 13, 15 or 17 can be used for talking to friends on other boats – but how do you know to which channel your friends are listening? The answer is that you wouldn't know, were it not that one channel, Channel 16, has been set aside for CALLING. It is also used for Distress and Urgency messages. Hence it is important to keep any operation on Channel 16 to a minimum.

The VHF GMDSS HANDBOOK

Table 1: INTERNATIONAL CHANNELS (Basic)

Channel number	Ship station frequency	Shore station frequency	Function of channel
00	156.000	156.000	HM Coastguard – Private Channel
01	156.050	160.650	
02	156.100	160.700	
03	156.150	160.750	Public Correspondence & Port Operations (two frequency)
04	156.200	160.800	
05	156.250	160.850	
06	156.300	–	Inter-ship *only*
07	156.350	160.950	Public Correspondence & Port Operations
08	156.400	–	Inter-ship *only*
09	156.450	156.450	Inter-ship & Port Operations
10	156.500	156.500	
11	156.550	156.550	Port Operations *only*
12	156.600	156.600	
13	156.650	156.650	Inter-ship & Port Operations
14	156.700	156.700	Port Operations *only*
15	156.750	156.750	Inter-ship & Port Operations
16	156.800	156.800	Distress, Urgency & Calling
17	156.850	156.850	Inter-ship & Port Operations
18	156.900	161.500	
19	156.950	161.550	
20	157.000	161.600	Port Operations *only* (two frequency)
21	157.050	161.650	
22	157.100	161.700	
23	157.150	161.750	
24	157.200	161.800	
25	157.250	161.850	Public Correspondence *only* (two frequency)
26	157.300	161.900	
27	157.350	161.950	
28	157.400	162.000	

Channel 16 acts as a central 'meeting place' or 'clearing house' where everybody first meets everybody else before changing to an appropriate working channel, as the others are called. The usual procedure for vessels, therefore, is to leave the set switched on and listen to Ch. 16 all the time you are on board. If anyone wants to talk to you *they will first call on Ch. 16*. Also, should a sudden emergency arise, Ch. 16 is the channel on which the distress or urgency message will be sent.

Although this is the *usual* procedure, there are now an increasing number of exceptions and alternatives to the general rule of making the initial call on Ch. 16. The precise procedure for calling specific radio stations is given in Chapters 7–12 below.

Two- frequency channels

Look closely at the frequencies allocated to each channel (Table 1) and you will notice that the channels in the centre of the band between Ch. 8 and Ch. 17, for Inter-ship and Port Operations use, have been allocated a single frequency for both ship and shore stations. For example, if you call a harbourmaster on Ch. 16 and he says 'go to Ch. 12' both you and he switch to Channel 12 and speak to each other on 156.6 MHz. In other words, you are both talking, literally, 'on the same wavelength'.

However, channels at the low-frequency and high-frequency end of the band (1–5, 7 and 18–28) which are used principally for 'Public Correspondence' (telephone calls to the shore) have been allocated *two* frequencies: one for the ship and a completely different one for the shore station. For example, if you call Niton Radio for a telephone call on Ch. 04 (Coast Radio Stations are called directly on one of the working channels, not Ch. 16 – see Chapter 10) your transmission is made on 156.2 MHz. Niton Radio must therefore listen for you on 156.2 MHz, which he does, so he hears you. However, his transmission is made on 160.8 MHz – a completely different frequency. To receive his transmission, therefore, your *receiver* must be tuned to 160.8 MHz – which it is, so you hear him. In practice there is no need to worry about this when making such a call, as the separate transmit and receive frequencies are automatically sorted out within your set when you select the appropriate channel number.

It is essential that you are aware that this situation exists, though, because on the two frequency channels you can talk *only* to the appropriate shore station; it is *technically impossible* to talk to other vessels. This confuses unqualified operators who mistakenly believe that the marine VHF band operates as a sort of 'maritime CB' which merely involves selecting a channel which no-one else is using. If the chosen channel is a two-frequency one, they will be disappointed. For example, in British waters you will rarely hear anything on Ch. 03; few British shore stations use this channel so it always seems to be available for use. It is *not*, because it is a two-frequency channel. So if you hear a vessel call another on Ch. 16 and suggest switching to Ch. 03, you know that they are unqualified. One station will make a transmission on 156.15 MHz but the other will be listening 160.75 MHz. They will *never* hear each other! (Q).

Below: Use the exclusive inter-ship channels 06, 08, 72 and 77 when speaking to other vessels

The VHF GMDSS HANDBOOK

You can talk to the appropriate shore station on a two-frequency channel because their transmitting and receiving frequencies are the opposite to yours: they are listening on your transmitting frequency and transmitting on your listening frequency. But every other vessel is the *same* as yours. On Ch. 03, for example, all vessels transmit on 156.15 MHz but listen on 160.75 MHz. *This cannot be altered.*

Channels to avoid

Although it is technically possible to talk to another vessel on any single-frequency channel, there are a number which should not be used. Ch. 11, 12 and 14 are reserved *exclusively* for harbour authorities. Do not use Ch. 13 in the Portsmouth, Plymouth and Clyde areas as the Queen's Harbourmasters at all British naval bases use this channel for talking to warships. Similarly, avoid Ch. 09, especially in the vicinity of major ports, as it is used by UK pilot vessels and harbour tugs. In the UK, Ch. 10 is used for pollution control actvities. **Voice operation is prohibited on Channel 70.** This is now reserved exclusively for Distress Alerts and routine calls by Digital Selective Calling (DSC).

For inter-ship operation, it is best to stick to the exclusive inter-ship channels of which there are now only four: 06, 08, 72 and 77.

Interleaved channels

By the early 1970s, the great increase in the use of marine VHF was causing congestion on the 28 channels available. More channels were needed but the IMM Band limits could not be extended. Instead, the width of each channel was reduced.

Reference to Table 1 shows that the channels were originally 50 kHz wide and were spaced at 50 kHz intervals. For example: Ch. 01 is centred on 156.05 MHz and Ch. 02 on 156.1, which is 50 kHz higher. On 1st January 1972, the width of all marine VHF channels was decreased from 50 kHz to the present 25 kHz. Thus a gap, also 25 kHz wide appeared between each of the original channels. In this way, the number of channels was doubled overnight without extending the limits and all marine VHF equipment was altered to suit. However, this caused a small problem with the numbering system. Rather than scrap the original system and start afresh by renumbering the channels 1–55, the original numbers were retained on the original channels. The interleaved channels, as they are known, were given completely fresh numbers. Unfortunately, numbers 29–59 had already been allocated to other services so the first number available was 60. This gave rise to the rather odd present situation seen in Table 2. This confuses the uninitiated who buy what is advertised as a 55 channel marine VHF set and are delighted to find that it actually goes up to Ch. 88. Their initial delight turns to dismay when they find that Channels 29 to 59 seem to be missing!

Private channels

Look closely at Table 2 and you will see that a large number of *frequencies* are missing. The highest transmitting frequency (Ch. 88) is 157.425 MHz but the lowest recieving frequency in column 3 (Ch. 60) is 160.625 MHz.

These 'missing' frequencies are also divided into channels for *private* use. Every qualified marine radiotelephone operator is fully authorised to use any of the international channels in Table 2 but, in addition, every maritime government may authorise its nationals to operate on one or more of these 'missing' frequencies on a private and exclusive basis. These private channels, as they are known, are allocated to corporate bodies such as ferry companies, towing and salvage companies.

The British Government has allocated two of the private channels for the use of British yacht clubs. The first, on 157.85 MHz and called Channel 'M', is also shared with British marinas. If your set cannot display 'M', you may have to select '37' or 'P1' or press a dedicated button. In 1989, a second private channel on 161.425 MHz, called 'M2', was released for British yacht clubs *only* together with International Channel 80 for British marinas, boatyards, repairers and chandlers *only*. It should be noted that, with one exception, British yacht clubs and marinas are not licensed for Ch. 16 and must, therefore, be *called directly* on the working channel.

As a point of interest, the private channels also occupy the range of frequencies from 162.05 MHz to 174.00 MHz.

The VHF GMDSS HANDBOOK

Table 2: INTERNATIONAL CHANNELS (Interleaved)

Channel number	Ship station frequency	Shore station frequency	Function of channel
00	156.000	156.000	HM Coastguard – Private Channel
60	156.025	160.625	
01	156.050	160.650	
61	156.075	160.675	
02	156.100	160.700	
62	156.125	160.725	
03	156.150	160.750	Public Correspondence & Port Operations (two frequency)
63	156.175	160.775	
04	156.200	160.800	
64	156.225	160.825	
05	156.250	160.850	
65	156.275	160.875	
06	156.300	–	Inter-ship *only*
66	156.325	166.925	Public Correspondence & Port Operations
07	156.350	160.950	
67	156.375	156.375	HM Coastguard Primary Safety Channel
08	156.400	–	Inter-ship *only*
68	156.425	156.425	Port Operations *only*
09	156.450	156.450	
69	156.475	156.475	Inter-ship & Port Operations (Ch. 10 UK Pollution Control)
10	156.500	156.500	
70	156.525	156.525	DSC Distress Alert and Routine Call
11	156.550	156.550	
71	156.575	156.575	Port Operations *only*
12	156.600	156.600	
72	156.625	–	Inter-ship *only*
13	156.650	156.650	Inter-ship & Port Operations (Ch. 73 UK Secondary Safety Channel)
73	156.675	156.675	
14	156.700	156.700	Port Operations *only*

The VHF GMDSS HANDBOOK

Table 2: INTERNATIONAL CHANNELS (Interleaved) *continued*

Channel number	Ship station frequency	Shore station frequency	Function of channel
74	156.725	156.725	Port Operations *only*
15	156.750	156.750	Inter-ship & Port Operations [REDUCED POWER]
75	–	–	GUARD BAND 156.7625–156.7875 MHZ
16	156.800	156.800	Distress, Urgency & Calling *only*
76	156.825	156.825	Direct-Printing Telegraphy for Distress & Safety
17	156.850	156.850	Inter-ship & Port Operations [REDUCED POWER]
77	156.875	–	Inter-ship *only*
18	156.900	161.500	Port Operations *only*
78	156.925	161.525	Public Correspondence & Port Operations
19	156.950	161.550	
79	156.975	161.575	
20	157.000	161.600	Port Operations *only* (two frequency)
80	157.025	161.625	
21	157.050	161.650	
81	157.075	161.675	Public Correspondence & Port Operations
22	157.100	161.700	Port Operations *only*
82	157.125	161.725	Public Correspondence & Port Operations
23	157.150	161.750	
83	157.175	161.775	Public Correspondence *only* (two frequency)
24	157.200	161.800	
84	157.225	161.825	Public Correspondence & Port Operations
25	157.250	161.850	
85	157.275	161.875	
26	157.300	161.900	
86	157.325	161.925	Public Correspondence *only* (two frequency)
27	157.350	161.950	
87	157.375	161.975	
28	157.400	162.000	
88	157.425	162.025	

4
WHO CAN YOU TALK TO?

Other ships

Most vessels at sea are fitted with VHF – even some sailboards! You may talk to other vessels *on matters of ship's business* quite freely at any time. Marine VHF must not be used for idle chit-chat; if that is what you wish to do, use CB (Q).

The usual procedure is to call the other vessel by name on Ch. 16, then switch to one of the designated *inter-ship* channels by mutual agreement. The full procedure is given in Chapter 7. Preferably, one of the *exclusive inter-ship* channels should be used: Ch. 6, 8, 72 or 77. Note these particularly, as it is only with inter-ship operation that *you* may have to make the choice!

It is quite permissible to contact another vessel directly on one of the inter-ship working channels, by prior arrangement, without first calling on Ch. 16 (Q). (Anything which reduces congestion on Ch. 16 has to be encouraged!)

Establishing contact with an unidentified vessel in sight can present a problem. The best way is to first establish communication by flashing "**Y, Y**" by light using the procedure on page 104 of the *International Code of Signals*, 1987, published by the International Maritime Organisation. Once

Below: When contacting an unidentified vessel, establish contact by flashing 'Y, Y' by light

Above: Harbourmasters can provide valuable information, but contact procedure must be checked

communication has been established (you now know the other vessel's call-sign and he knows your's), you can call the vessel on Ch. 16 in the usual way.

NOTE: Inter-ship communication is *not permitted* while the ship is within, or within one mile of, any port, harbour, dock or anchorage in the territorial waters of any country – except in the case of distress, emergency involving danger to life or to navigation, for the purposes of safe navigation, or in the Port Operations Services. Inter-ship communication *is* permitted on inland waterways provided that both vessels are under way.

Coastguards

Coastguards are concerned for *your safety* and you may talk to them on any aspect of safety quite freely at any time. All the major stations listen on Ch. 16, 24 hours a day, 365 days a year. Their names are the names of their location plus the word 'Coastguard', thus: 'Solent Coastguard', 'Dover Coastguard', etc.

After establishing contact on Ch. 16, British Coastguards will tell you to switch to Ch. 67 (Q).

The procedure for calling the Coastguard is given in Chapter 8.

All the major British Coastguard Stations are now fitted with direction-finding equipment which operates on Ch. 16. In case of difficulty, they may be able to give an indicated bearing upon request.

Harbourmasters

Most harbours throughout the world use VHF for providing harbour information. Many of them may be called, initially, on Ch. 16 and you will then be asked to switch to an appropriate Port Operations channel. The most popular are Channels 12 and 14, followed by 11 and 13, but many others are used. However, there are no general rules regarding the calling channel, the hours of watch or even the name to call! Some harbours listen only on their working channel, some listen only on Ch. 16. Others listen on both but prefer to be called directly on their working channel. Only the major ports maintain 24-hour watch; smaller harbours decide for themselves when they switch their sets on and off. There is no charge for communicating with a port or harbour.

Most harbours are addressed by their names *plus* the words 'Harbour Radio', thus: 'Cowes

The VHF GMDSS HANDBOOK

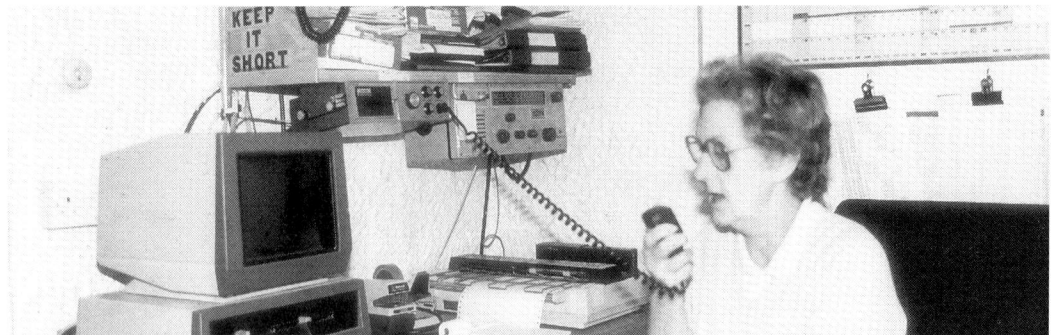

Above: British marinas and yacht clubs are only licensed to use Channels 80, M2 or M

Harbour Radio', 'Langstone Harbour Radio', etc. Many of the major ports, however, are addressed differently. Examples are: 'Mersey Radio' (for Liverpool), 'Long Room Port Control' (for Plymouth), 'VTS' (for Southampton). All British naval bases, which use Channel 13, are addressed as 'Queen's Harbourmaster' or 'QHM'. For full information, consult the *Admiralty List of Radio Signals*, Vol. 6. Available from all Admiralty chart agents. Part 1 covers Northern Europe. Part 2 covers the Mediterranean, Africa and most of Asia. Part 3 covers the rest of the World. Abridged information is published in the yachting almanacs.

Marinas and yacht clubs

For many years, British marinas and yacht clubs have shared the private frequency of 157.85 MHz called Channel 'M'. This is a private channel which may *only* be used for yacht club or marina business in Britain, Channel Islands and Isle of Man. Only *British yachts* may use this channel; the necessary authority is usually included with the yacht's Ship's Licence at no extra charge. On sets which cannot display the letter 'M', the number '37' or 'P1' may have to be selected. Some sets have a special button for Ch. M; consult the owner's handbook.

As Ch. M is not available to foreign yachts, the RA released Channel 80 for use by British *marinas* in 1989 and for *boatyards, repairers and chandlers* in 1998. This is now the primary channel for British Marinas. Also in 1989, the RA released a second private channel 'M2', for use by British *yacht clubs*. The frequency is 161.425 MHz and is intended for race and regatta control. In some older sets, it can be programmed as 'P2'.

With one exception, British marinas and yacht clubs are *not authorised* for Ch. 16 and therefore *must be called directly* on Channels 80, M2 or M as appropriate. This procedure also applies to marinas on the tidal Thames and on all inland waterways.

The exception is Brighton Marina which may be addressed as 'Brighton Control' on Ch. 16 if Ch. M is not available on the yacht. On Ch. M, it is addressed as 'Brighton Marina'.

Most French and Spanish marinas and yacht harbours use Ch. 09 *only*.

Apart from the exceptions given above, details of British yacht clubs and marinas are not given in the *Admiralty List of Radio Signals*. Details *are* given in the yachting almanacs but only when this is provided by the club or marina. British marinas are usually addressed by their name; yacht clubs by name *plus* the word 'Base', thus: 'Parkstone Base', 'Bosham Base', etc. There is no charge for communicating with a yacht club or marina.

Coast Radio Stations

These are usually operated by the telephone company/authority of the country concerned and form shore-side 'telephone exchanges' which connect ships with the worldwide telephone system. They also accept dictated telegrams. All Coast Radio Stations throughout the world speak English. They are addressed by the name of their location *plus* the word 'Radio', thus: 'Niton Radio', 'Hastings Radio', etc.

Throughout Northern Europe, Coast Radio Stations do not listen on Ch. 16. Modern practice is to call directly on one of the station's working channels. Details of these and much more can be found in the *Admirality List of Radio Signals*, Vol. 1, of which Part 1 covers Europe, Africa and most of Asia, and Part 2 covers the rest of the World. Abridged information is also published in the yachting almanacs (Q). An indication that a British channel is engaged is given by hearing speech or a succession of 'pips' rather like the Greenwich 'pips' but slower. If neither speech nor 'pips' are heard, the channel is available for use (Q). If a Coast Radio Station does not answer immediately, the 'pips' will be heard as soon as the microphone switch is released – provided that your transmission lasts at least ten seconds. This indicates that your call has been received and a reply will be made soon. The full procedure for obtaining a ship-to-shore telephone call is given in Chapter 10.

Payment

Although there is no charge for calling a Coast Radio Station, a charge is made for successful connection to the land-line. Throughout the UK there is only one standard charge – there are no peak rates or cheap rates.

Once communication with a Coast Radio Station has been established, the method of payment must be stated before the connection is made. This is done by stating your Accounting Authority Identification Code (AAIC) or 'Account Code' spoken as 'GOLF, BRAVO, (NUMERAL), (NUMERAL)' (Q).

There are several methods of payment. The cheapest is the 'YTD' system. This stands for 'Yacht Telephone Debit' and is a concession by British Telecom to allow the cost of a radiotelephone call to be directly debited to a home or office telephone account. To use this method, tell the Coast Radio Station 'My account code is YTD' plus the number to which you wish the call to be charged. This must include the area code. For example: 'My account code is YTD 0181–234–5678'. The net cost of the call will then appear on the next quarterly bill for this number. In this way a crew-member or guest can make a telephone call and have it charged to his or her home telephone number.

The YTD system can *only* be used for calls from *B*ritish yachts to *B*ritish telephones via *B*ritish Coast Radio Stations on VHF. If the call does not fulfil these requirements, an International Account Code Number must be given.

When an application is made for a Ship's Radio Licence, an authorised accounting company must be appointed (from a selection offered) to handle the vessel's radio accounts. For occasional telephone calls, BT subscribers are advised to appoint BT Aeronautical & Maritime Billing, whose AAIC is 'GB14'. When this code is given to a Coast Radio Station *anywhere in the World*, British Telecom send an entirely separate bill to the name and address which was given in paragraph 2 of the Ship's Licence application. A small handling charge will be added to the bill so, although 'GB14' *can* be used for British calls, the 'YTD' method is cheaper.

For frequent calls or non-BT subscribers, CI Maritime Services of Guernsey (GB19) levy a small returnable deposit to open an account but do not charge VAT!

'Transferred Charge' (reversed charge) or 'collect' calls can be made provided the recipient agrees to accept the charge, which is the cost of the call plus an extra charge.

In theory only, calls via British Coast Radio Stations can be charged to a credit card – however this method is not advised. The card number may be overheard by an unscrupulous person who could use it for their own calls!

Coast Radio Station broadcasts

In addition to *communicating* with individual vessels, Coast Radio Stations and many Coastguard stations broadcast vital information to shipping in general.

Gale warnings

The appropriate Coast Radio Station broadcasts gale/storm warnings immediately upon receipt from the Meteorological Office (at Bracknell in the case of Britain). The initial announcement is given on Ch. 16 in the form:

    SÉCURITÉ, SÉCURITÉ, SÉCURITÉ
    ALL SHIPS, ALL SHIPS, ALL SHIPS
    THIS IS (somewhere) RADIO,
        (somewhere) RADIO
    FOR GALE WARNING, LISTEN
    (frequency and channel number)

The stated channel number should then be selected, and the gale warning is read out on that channel after a short interval. It is repeated on a six-hourly schedule until it is cancelled when the cancellation is also broadcast. For full information consult the *Admiralty List of Radio Signals*, Vol. 3.

As the major Coast Radio Stations broadcast simultaneously on the Medium Wave Band (around 2 MHz) as well as VHF, the initial announcement includes the Medium Wave frequency. This cannot be received on a marine VHF R/T set, so it should be ignored.

NOTE: The French word SÉCURITÉ (pronounced SAY–CURE–E–TAY), meaning 'safety', precedes all broadcasts of navigational importance to alert ships to this fact.

Navigational warnings

Warnings of navigational hazards of a temporary nature such as wrecks, lights extinguished or radio beacons inoperative are broadcast by the appropriate British Coast Radio Station every four hours on its broadcast working channel following an initial announcement on Ch. 16. All navigational warnings are preceded by the word 'SÉCURITÉ'.

British Coastguard Stations also broadcast Navigational Warnings on Ch. 67 following an announcement on Ch. 16. In the Dover Strait, Dover Coastguard broadcasts navigational information on Ch. 11 every H+40 (and H+55 in poor visibility). The French coastguard service for the English Channel (C.R.0.S.S.M.A.) broadcasts navigational information in English and French from Cap Gris Nez, Jobourg (Cherbourg Peninsula) and Ushant on Ch. 11 every half-hour (quarter-hour in poor visibility). For full information consult the *Admiralty List of Radio Signals*, Vol. 6, Part 1.

Weather forecasts

British Coast Radio Stations broadcast the General Synopsis and Forecast for their own Sea Area and those immediately adjacent twice daily at fixed times in the morning and evening (varies with station). The forecast goes out on the main broadcast channel following an announcement on Ch. 16. Immediately after the first reading, the information is repeated at dictation speed. Thus, not only is the information given at a more civilised hour than the Shipping Forecast on Radio 4, it is also much easier to copy by hand.

Jersey Radio broadcasts the General Synopsis and Forecast for the Channel Islands on request and five times a day at 0645, 0745, 1245, 1845 and 2245 on Chs 25 and 82. Continental Coast Radio Stations also frequently broadcast gale warnings and weather forecasts. For full information, consult the *Admiralty List of Radio Signals*, Vol. 3 which covers

the world. Note that although most Coast Radio Stations throughout the world broadcast gale warnings and weather messages in English, the French do not. All Met. broadcasts from French Coast Radio Stations and those of their territories and Tunisia are in the French language only.

In United States waters, the US Coastguard Stations frequently broadcast storm warnings and weather forecasts on the special Channel 22CG. In this case, Ch. 22 is used as a *Single-frequency channel* with the shore station transmitting on the ship frequency of 157.1 MHz. European sets usually need modification to receive this frequency, or it may be programmed as a private channel if the set has the capacity. In highly populated US coastal areas, taped weather messages are broadcast every four to six minutes on one of three private channel frequencies: 162.55 MHz (called 'WX1'), 162.4 MHZ ('WX2') and 162.475 MHz ('WX3'); the messages are updated every two to three hours. These frequencies may also be programmed into some European sets; with the very latest sets they can be built in by the manufacturer.

In Canada, a similar continuous broadcast of weather information is made by the Coast Radio Stations on Chs 21 or 83 according to location. Full information on these and USCG stations can be found in ALRS, Vol.3.

Traffic lists

Coast Radio Stations will accept telephone calls *from* the shore *to* a ship. The charge is the same either way. The procedure is to dial Freephone, 0800–378389 giving the boat's name, call-sign, voyage details and other information. These details are then passed to the appropriate Coast Radio Station which then calls the boat on Ch. 16. If answered immediately, the CRS dials the caller and makes the connection.

In the case of no reply, the boat's name and call-sign are then added to the Coast Radio Station's Traffic List. British stations broadcast their Traffic List 10 times between 0133 and 2233 (varies with station) together with other broadcasts such as Weather Forecasts, etc. The schedule of Traffic Lists is given in ALRS, Vol. 1.

Should you hear your vessel's name on a Traffic List, call the Coast Radio Station *on one of its working channels* and say 'What have you for me ?' (Q). The radio officer will then dial the caller's number and make the connection. The original caller still pays. The boat's name and call-sign are retained on the Traffic List for 24 hours; after that it is lost.

The VHF GMDSS HANDBOOK

Above: While it is likely the Coastguards in the UK will maintain a listening watch on Channel 16 into the 21st Century, this won't always be the case elsewhere!

Distress Signals

Until 1999, distress signals may be broadcast (i.e. not addressed to any particular station) by R/T on Ch. 16. Although broadcast, distress signals will usually be acknowledged by a Coastguard station who will subsequently control the situation. However, the procedure for obtaining help at sea will be completely different from lst February, 1999 with the final implementation of the GLOBAL MARITIME DISTRESS and SAFETY SYSTEM (GMDSS). From that date, distress alerts will be broadcast by a Digital Selective Calling (DSC) signal on Ch. 70. This will be generated automatically by an electronic circuit (within the set or, perhaps, an external accessory) and actuated simply by pressing a 'Distress' button. Subsequent distress communications will still be conducted on Ch. 16 as presently.

It is vital for leisure sailors to appreciate that, under GMDSS, there is no provision for Coastguards to maintain a listening watch on Ch. 16. Although it likely that British Coastguards will continue to maintain a watch on Ch. 16 for some time into the 2lst Century, it is probable that Continental Coastguards will not! When considering the purchase of a VHF set in the future, therefore, ensure that it conforms to 'Class D'.

5

SOME BASIC TECHNICALITIES

Although no technical knowledge is required for any of the VHF Certificates, as with the Driving Test, it is of great benefit to have a rough idea of some of the basic principles.

SIMPLEX

The vast majority of radio communication stations operate in what is known as the Simplex mode. Inside, the box of electronics is divided into two parts. One is the *transmitter*, to which the microphone is connected for talking. The other is the *receiver*, to which is connected the loudspeaker or earpiece for listening. Both share the same power supply (usually a 12V battery) and both require an aerial or antenna. The same aerial can be used for transmission *or* reception – but *not both at the same time*. So with a Simplex installation, you have to decide whether to use your set for talking *or* listening since you cannot do both at once!

The change-over is achieved with a spring-loaded *press-to-talk* (PTT) switch fitted to the hand microphone or concealed in the handle of a telephone-type handset. With the set switched on and the PTT switch in the released position, the set is in the listening mode. To make a transmission, the PTT switch must be pressed and held pressed all the time you are talking. The PTT switch must then be released to listen. This may sound inconvenient but, with a few minutes' practice, the 'press-to-talk, release-to-listen' habit is quickly acquired.

As the name implies, the Simplex system keeps the equipment simple and inexpensive, although it is not quite as easy as operating a normal domestic telephone. It also requires the co-operation of the other person, since they must wait for you to finish speaking before they can reply. Having said something, therefore, you say 'OVER' just before releasing the PTT switch to indicate to the other person that you are about to pass the channel 'over' to them.

This works well between two people who are versed in this one-way-at-a-time system, but it can

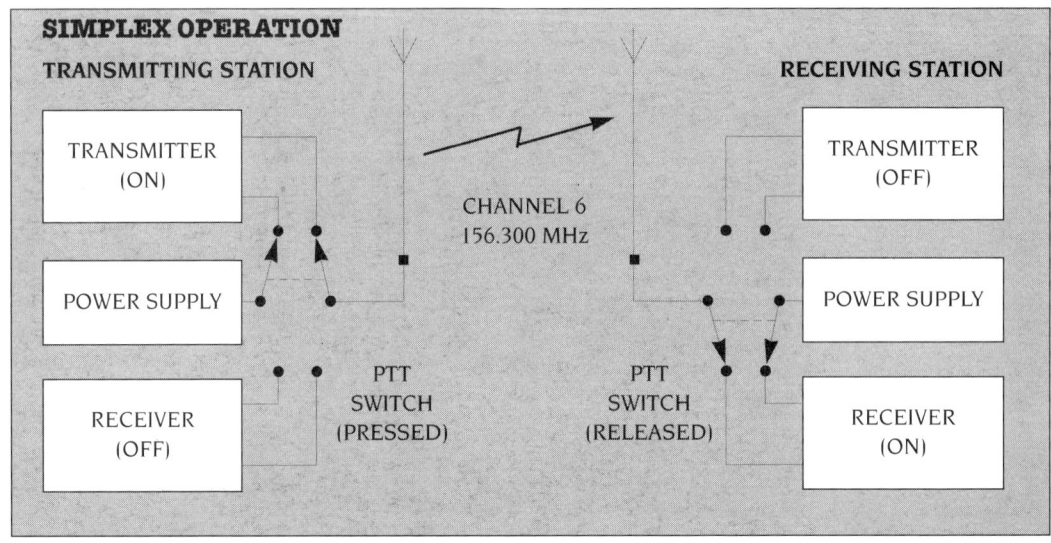

The VHF GMDSS HANDBOOK

be a little daunting for a completely uninitiated person such as a ship's passenger telephoning a person ashore.

DUPLEX

The Duplex system allows for simultaneous transmission and reception so that an ordinary two-way conversation may be carried out between a vessel and a telephone ashore.

Unless an electronic switching device (duplexer) is used, two separate aerials are needed: one for the transmitter and a second for the receiver. To operate on the same frequency these must be widely separated – if they were not, most of the transmitted power would be picked up by the receiving aerial and probably damage the receiver. On a small ship, however, the two aerials can never be very far apart.

The problem is reduced by arranging for the two stations to transmit on two completely different, widely-separated frequencies. The ship's receiver is tuned to the signal from the shore transmitter and, similarly, the shore receiver is tuned to pick up the signal from the ship. Even with two different frequencies, a good separation must still be maintained between the transmitting and receiving aerials, otherwise the transmission may overcome the frequency separation!

All Coast Radio Stations are equipped for Duplex operation; this is why all the public correspondence channels are two-frequency channels. It should now be clear why the two frequency channels cannot possibly be used for inter-ship communication. On Ch. 04, for example, *all* ships transmit on 156.2 MHz and listen on 160.8 MHz. Not only can you not hear yourself (there is no need), no other ship can hear you either! (Q).

'SEMI-DUPLEX'

The average yacht's Simplex R/T set may be used for telephone calls to the shore via Coast Radio Stations provided the set has some, if not all, of the two-frequency channels and assuming the caller on the yacht has the co-operation of the person ashore. Although the latter can speak and listen simultaneously, the person afloat cannot: the person ashore should be advised of this beforehand, for this will save a good deal of confusion, explanation, time and money if you need to make a call.

RANGE

VHF radio waves travel in straight lines, just like rays of light radiating from a lighthouse. In the same way, VHF range is restricted by the curvature of the earth. Provided sufficient power is transmitted (and 25 watts (Q) is about ten times the minimum required), the aerial-to-aerial range is determined by the height of the respective aerials above sea level (asl).

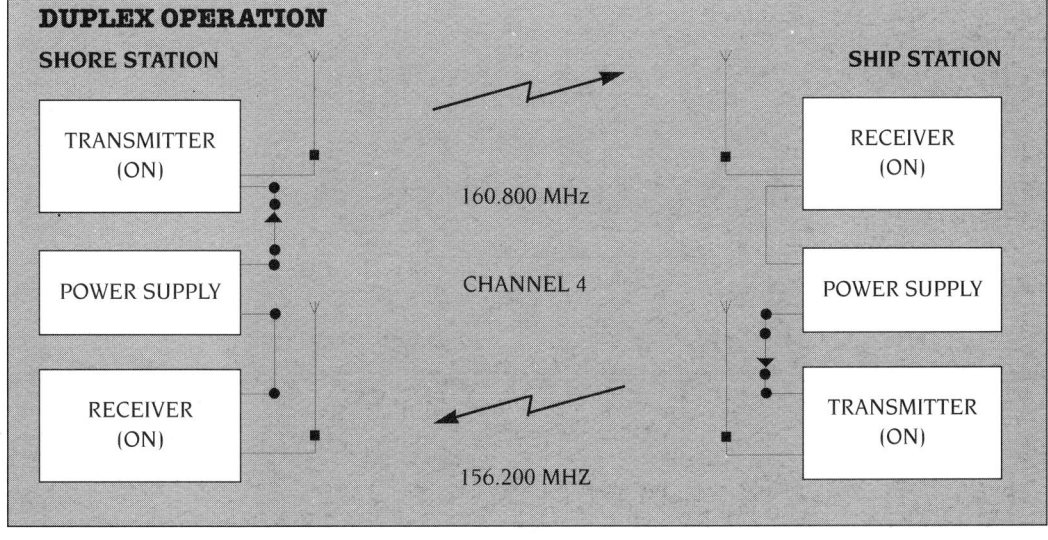

As the VHF waves behaves like an invisible beam of light, the tables used for determining the extreme range of a light at sea can also be used for determining the range of a VHF signal. In *Macmillan's* the extreme range is referred to as the '*Distance Off Lights Just Seen or Dipping*'. To use the table in this way, simply enter the height of one aerial as the 'Height of Light' and that of the other aerial as the 'Height of Eye' (it does not matter in which order the aerials are taken). The junction of the two columns gives the range in nautical miles. As the VHF wave is refracted (bent) towards the Earth slightly more than light, the figure thus obtained may be increased by ten per cent for maximum accuracy. However, accepting the figure as quoted – often referred to as 'line-of-sight' range – gives a little to spare for safety.

In the absence of tables, the range under any circumstances is easily calculated if you have a pocket calculator. For *each* aerial, the range in nautical miles to the 'radio horizon' (slightly greater than the visible horizon) is given by the formula $R = 2.25\sqrt{h}$, where h is the height of the *centre* of the aerial above sea level in metres. The two ranges thus obtained are then added to give the total aerial-to-aerial range. For example, two cabin cruisers or small fishing boats, both with aerials at 3m (10ft) asl would have a horizon range as follows:

$R = 2.25\sqrt{h}$
$R = 2.25\sqrt{3}$
$R = 2.25 \times 1.73$
$R = 3.9$ nautical miles for *each* aerial

Thus, their boat-to-boat range would be 3.9 + 3.9 = 7.8 nautical miles *only*. No increase in power would increase the range; remember that the VHF signal is like a light at sea – no matter how powerful the light, it is invisible below the horizon!

The 'rag-and-stick' sailor with an aerial at the top of his mast fares much better. Even a small yacht's aerial would be 10m asl, giving a horizon-range of just over seven nautical miles or 14.25 miles boat-to-boat. This is nearly twice that between two power boats. *These are typical boat-to-boat ranges* (Q).

Much greater distances can be achieved by all vessels to shore stations which can put their aerials on tall masts on high ground. For example, a small yacht with an aerial at 10m asl can communicate with a shore station with an aerial at 100m asl at the following range:

$R = 2.25\sqrt{h} + 2.25\sqrt{h}$
$R = 2.25\sqrt{10} + 2.25\sqrt{100}$
$R = 2.25 \times 3.16 + 2.25 \times 10$
$R = 7.12 + 22.5$
$R = 29.62$
say 30 nautical miles (Q)

Capture effect

Marine VHF uses the system of Frequency Modulation (FM). The end result is much the same as that produced by Amplitude Modulation (AM), as used on the Medium Wave and Short Wave Bands, but an important feature of FM is that the receiver can only reproduce one signal at any one time. If several signals are being received at once, the FM receiver 'locks-on' to the strongest signal and reproduces it to the complete exclusion of all others (Q). As a result, you never hear the unintelligible jumble of voices typical of the Medium Wave Band at night when Continental stations are strong. This, plus the fact that the 25 watts allowed is grossly in excess of the minimum required to achieve the fairly short range, ensures that marine VHF signals tend to be mostly 'loud and clear' or nothing!

It is for this reason that a low-power (1 W) switch is fitted to the transmitter section of all marine VHF sets with an output power in excess of 1 W. To avoid causing unnecessary interference to other stations, the minimum amount of power should always be used (Q). This benefits the users also, as it saves battery power. (The average marine VHF draws about 0.33 amps on receive, 1 amp on low-power transmit and 5–6 amps on high-power transmit).

Low power should always be used for the short-range inter-ship communication and for communicating with local harbour masters, yacht clubs, marinas, boatyards, repairers and chandlers. Full power will not be of any benefit, and will only cause interference to other stations some distance away (Q).

Low power will also suffice for routine calls when in sight of a Coastguard Station but any Distress or Urgency calls should, of course, be made on full power. Full power should also be used for telephone calls via Coast Radio Stations, who like to put a strong signal down the line.

Dual watch

Most modern marine VHF sets of the 'fully-synthesised' 55-channel type feature an extremely useful facility known as dual watch. This is a circuit which *effectively* (though not actually) enables the set to receive on two channels at the same time: Channel 16 (which has priority) and any one of the remaining 54 channels as selected.

The VHF GMDSS HANDBOOK

Above: A typical VHF (non-DSC) set for installation in a boat. Note the PTT switch on the microphone

In dual-watch mode, the receiver switches continuously (scans) between the channel selected and Ch. 16, spending most time on the selected channel. Should a transmission be received on the working channel, it will be heard with fair clarity although, since the receiver will continue to scan Ch. 16 momentarily, the odd syllable may be missed. If this happens, the dual watch facility may be switched off to allow uninterrupted reception on the working channel.

If a signal is received on Ch. 16 with dual watch in operation, the receiver will 'lock-on' to it *and remain on Ch. 16 during the whole transmission.* Thus 100 per cent of transmissions on Ch. 16 and 90 per cent of transmissions on the selected channel will be received.

This feature is especially useful if private calling arrangements have been made. Although it is usual to make the initial contact with other vessels on Ch. 16, it is quite acceptable to call directly on an inter-ship working channel (06, 08, 72 or 77) at a pre-arranged time by private agreement (Q). By using dual watch, a listening watch may then be kept on. Ch. 16 as well as the pre-arranged channel (Q).

Always switch off dual watch before transmitting. Some sets transmit on Ch. 16 when in dual watch mode, some on the working channel; the most up-to-date sets will not transmit at all.

Squelch

Because of the enormous amount of signal amplification which occurs in a VHF receiver, a great deal of noise (hash) is generated which can be very objectionable between signals. When a signal is received, it over-rides the noise so only the signal is heard but, between signals, you get a loud hiss. The squelch circuit combats this by muting the loudspeaker in the absence of a signal so the hiss is not heard. As soon as a signal is received, however, the muting system is overridden, enabling the signal to be heard (Q).

For best results the squelch control should be turned off (fully anti-clockwise) then advanced slowly until the hiss *just* stops. If it is advanced too far, the sensitivity of the receiver will be reduced so much that weak signals may be missed. The setting will not vary as the channels or volume control are adjusted, but it may vary slightly with fluctuations in battery voltage and it may get knocked, so check the setting frequently (O).

Selcall

The Selcall (Selective Calling) system enables a Coast Radio Station to call an individual ship, not by calling out the name on Ch. 16 but by playing a short electronically-generated tune (Q). Every ship participating in this scheme is allocated a 'signature tune' which is decoded in the receiver. The receiver then flashes lights and/or sounds a warning, and also records the fact that the set has been called. Some decoders also record the indentification of the calling station. Although little used for yachts, the system could be useful if you wish to leave your yacht for a short period. Upon your return, the set would indicate whether or not it had been called in the meantime without you having to wait nearly two hours for the next Traffic List. Selcall becomes obsolete for R/T in 1999.

Test call

An extremely useful and inexpensive optional extra, easily fitted to all sets, is an output test meter which gives a direct indication of the transmitted output power without having to rely on a friend or Coastguard for a 'radio check'. The more sophisticated devices also indicate the efficiency of the aerial and cable, enabling the performance of the whole transmitter/aerial system to be measured every time the transmit switch is pressed. Very reassuring.

It takes ten seconds to complete the test and this is the maximum time allowed if you are not in communication with someone (Q). If you are making a test transmission without addressing another station, you must still announce your identification in the usual way, even though you are not addressing anyone.

6

PRIORITY OF SIGNALS ON CHANNEL 16

Until 1999, initial contact with other stations is usually made on Ch. 16 (but see Chapter 13). As Ch. 16 is also used for Distress and other emergency calls, an order of priorities has been established.

Distress signals

Top priority for the use of Ch. 16 is given to distress signals. A distress situation is indicated by the use of the word 'MAYDAY'. This indicates that the *vessel* (or aircraft) is in *grave* and *imminent* danger, that is to say *sinking* (or about to sink) or *on fire* (Q). Under these circumstances it might be thought that 'Mayday' is a silly word to say – but it is not what it seems. Although English is the international language of radio, all the procedural words, or *prowords* as they are termed, are French. So, if you get into difficulties, you cry for help in French! The French spell it *'m'aidez'* meaning literally, 'come to my aid' or in basic English 'HELP'!

The distress signal is very special for two reasons. It is a *broadcast* – it is not addressed to anyone in particular – and it is the only occasion on which *messages are exchanged* on Ch. 16. Consequently, whenever there is a distress situation, *radio silence is automatically imposed throughout the whole operation*. An announcement that normal or restricted working may be resumed on Ch. 16 will be made, on Ch. 16 (in French), by the station controlling distress communications. This may be the distressed vessel itself but is more likely to be a Coastguard Station or possibly a Coast Radio Station. For this reason *it is vital to keep listening on Channel 16* to keep abreast of the current situation. The procedure for a Mayday broadcast, which must be *authorised* (but not necessarily made) by the *master* of the vessel, is given in Chapter 11.

Urgent calls

All emergency calls apart from sinking or fire are prefixed by the prowords 'PAN-PAN' (Q). This again is derived from the French *en panne* meaning 'in difficulty' and takes second priority to distress. It *is not a cry for help*. It is simply a plea for radio priority: radio silence is observed during the message *and for three minutes after*. As it is not a broadcast, the 'PAN-PAN' message must be addressed to someone. Usually it will be addressed to a Coastguard Station but, if the vessel is out of range of the shore, the message may be addressed to 'All Ships'.

A variation, 'PAN-PAN MEDICO', is used to prefix calls to British Coast Radio Stations requesting urgent (and free) medical advice (Q). The procedure for obtaining medical advice via foreign Coast Radio Stations is given in the *Admiralty List of Radio Signals*, Vol. 1.

Short 'PAN-PAN' messages may be broadcast to 'All Ships' on Ch. 16, but a Coastguard Station will ask a vessel to switch to a working channel: Ch. 67 in Britain (Q).

Safety calls

Broadcast warnings of strong winds and navigational hazards by Coast Radio Stations and Coastguard Stations are prefixed by the proword 'SÉCURITÉ' (pronounced 'say-cure-ee-tay') which is French for 'safety'. As yachts do not broadcast, no procedure is given in these pages. However, if you wish to advise a Coastguard Station of a newly-found navigational hazard, the initial call could be preceded by the word 'SÉCURITÉ'.

Other priorities

In all, there are ten internationally-agreed priorities listed in the *Handbook for Marine Radio Communications* 2nd Ed (ISBN 1–85978–041–5) but the above are the most important.

Routine calls

Provided none of the above situations are in operation, routine calls may be made on Ch. 16. The procedure for calling different stations is given in the following pages.

7

SHIP-TO-SHIP COMMUNICATION

This is probably the most common use of the VHF/RT set. It is also the easiest procedure. As you will normally be calling someone you know or can see, it is usual to use vessel names rather than call-signs (Q). However, call-signs can be used if known in case of difficulty or duplication of names (Q). Many yachts now display the call-sign on the dodger for easy identification and this is strongly recommended. The author had the call-sign sewn onto the mainsail in two-foot high letters, and on the mizzen in 18-inch high letters; they could be seen from more than a mile away with good binoculars!

Although the usual procedure is to make the initial call on Ch. 16, it is quite permissible to call directly on one of the inter-ship channels (06, 08, 72 or 77) by prior arrangement (Q).

As an example, with yacht *Gaffer* calling another *Flash*, it will be assumed that the initial call is being made on Ch. 16 as this is the usual case. Having made quite sure that Ch. 16 is quiet, *Gaffer* presses the press-to-talk button:

Before switching channels, it is *vital* that the calling station confirms the working channel:

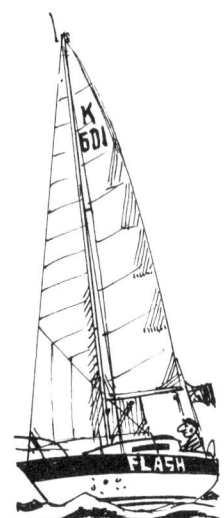
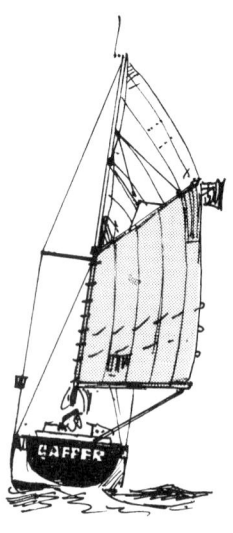
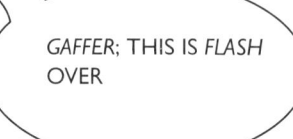

FLASH; THIS IS *GAFFER*
CHANNEL 8
OVER

Both vessels then switch (in this case) to Ch. 08. As the station *called* controls communication (Q), *Flash* re-opens the conversation, having established that the channel is not in use:

GAFFER; THIS IS *FLASH*
OVER

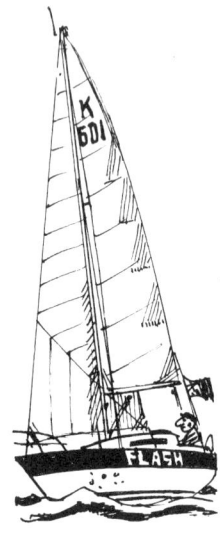
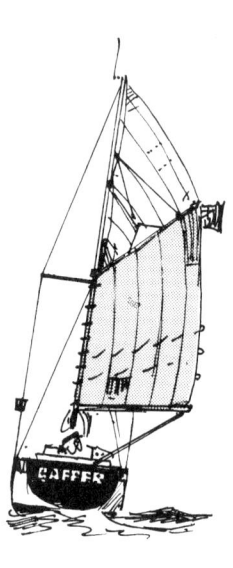

FLASH; THIS IS *GAFFER*
WHAT PORT ARE
  YOU AIMING FOR?
OVER

GAFFER; THIS IS *FLASH*
KEYHAVEN
OVER

The exchange can continue as long as *necessary* (Q) with each station saying 'Flash; this is Gaffer' or 'Gaffer; this is Flash' every time the microphone button is pressed (Q). At the end of the exchange of messages, the usual 'goodbye' pleasantries can be exchanged but, finally, the communication is closed as follows:

The VHF GMDSS HANDBOOK

Both stations then switch back to Ch.16, ready for any further calls.

Direct Calling
The initial call to another vessel may be made directly on an inter-ship channel (6, 8, 72 or 77) by private, prior arrangement (Q). In this case, the procedure is simplified to 'Flash; this is Gaffer, OVER' and the reply 'Gaffer; this is Flash. Go ahead, OVER'. On these occasions, the dual-watch facility (if fitted) proves invaluable as it enables a listening watch to be kept on any working channel as well as on Ch. 16 (Q).

Indistinct Reception
If a completely garbled message is received, a repetition may be requested by asking the sending station to **say again**. (*The word 'repeat' should not be used as it is used to emphasize a word or phrase.*) If part of a message is garbled, the sending station should be asked to **say again all after/all before/all between/word after/word before**, or **word between the part(s) received clearly** (Q). Repetitions should be preceded by the phrase '**I say again**' (Q).

Unidentified Caller
If the identification of a calling station is unclear, it should be addressed as 'station calling (your ID), this is (your ID), say again your ship's name/call-sign (Q).

Unclear Addressee
If you **think** that you are being called but are unsure, *wait until the call is repeated* (Q).

8

TO CALL COASTGUARDS

The prime task of Her Majesty's Coastguards is safety of life at sea but they also have a secondary task to co-ordinate pollution control operations. Contrary to popular belief, they are not now concerned about contraband and have not been since about 1860!

The Coastguards are a civil, uniformed service of the Maritime and Coastguard Agency (MCA) and operate a chain of strategically-placed Coastguard Stations all round the UK coastline. Although most have a view of the sea, visibility in UK waters is rarely good so great reliance is placed on VHF Radiotele-phony. So, although there are now many fewer Coastguards manning Lookouts (as they were called) than there were some years ago, the service is now very much better than it was. Only a few years ago, you were lucky if a Coastguard could *see* you three miles away; now he or she can *hear* you 30 miles away! For this reason it is now most important that everyone who goes to sea carries a VHF radiotelephone.

Coastguard Station Weather Broadcasts

Station	Time	Station	Time
Aberdeen	0320	Milford Haven	0335
Belfast	0305	Oban	0240
Brixham	0050	Pentland	0135
Clyde	0020	Portland	0220
Dover	0105	Shetland	0105
Falmouth	0140	Solent	0040
Forth	0205	Stornoway	0110
Holyhead	0235	Swansea	0005
Humber	0340	Thames	0010
Liverpool	0210	Tyne–Tees	0150
		Yarmouth	0040

The above are starting times – the broadcast is repeated every four hours following the time given

Weather forecasts

In addition to acting as communication centres, Coastguard Stations also regularly broadcast weather forecasts, strong wind warnings and local navigational warnings on Ch. 67 after a preliminary announcement on Ch. 16. The table on this page gives the starting time (local) for HMCG weather broadcasts. These are then repeated at four-hourly intervals, or two-hourly intervals if a gale warning or strong wind warning is in force.

Navigational help

For the Dover Strait, Dover Coastguard broadcasts a 'Channel Navigation Information Service' on Ch. 69 at H+40 during fine weather and again at H+55 during poor visibility, following an initial announcement on Ch. 16. Their French counterparts, Gris Nez Traffic, Jobourg Traffic and Ouessant Traffic (Ushant) do the same in English and French on Ch. 79.

Dover Coastguard controls radar stations at St Margaret's Bay and Dungeness. Vessels within range of these stations can ask for a range and bearing from either by calling Dover Coastguard on Ch.69.

Most UK Coastguard Stations can now give the *bearing* (but not *range*) of a station calling on Ch. 16. By 'crossing' two such bearings, say from Portland Coastguard and Solent Coastguard, a fair 'fix' can be obtained. The bearings have an accuracy of plus or minus 2 degrees at best, so the system is not highly accurate. In poor visibility after a long and hard trip across the Channel, though, such a fix will be far better than an estimated position or one obtained by dead reckoning. Although primarily intended for use in an emergency situation, the service is freely available for anyone who *needs* it. Obviously the Coastguards cannot cope with 5,000 yachtsmen in the Solent area each calling for a bearing every half-hour on a sunny afternoon, so the service must not be abused.

The VHF GMDSS HANDBOOK

Calling procedure for routine calls

The yacht Gaffer (call-sign MABC) approaching Torbay and wishing for a report of the sea conditions there, would call Brixham Coastguard on Ch.16:

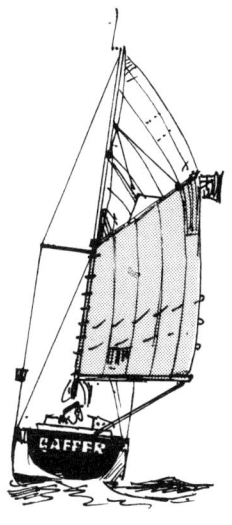

BRIXHAM COASTGUARD,
THIS IS MIKE ALPHA BRAVO CHARLIE,
MIKE ALPHA BRAVO CHARLIE
SAFETY MESSAGE; CHANNEL SIX SEVEN
OVER

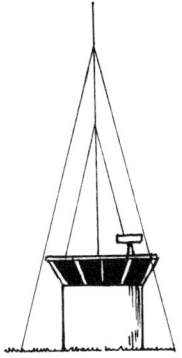

MIKE ALPHA BRAVO CHARLIE,
THIS IS BRIXHAM COASTGUARD,
CHANNEL SIX SEVEN AND STAND-BY
OVER

BRIXHAM COASTGUARD
THIS IS MIKE ALPHA BRAVO CHARLIE
CHANNEL SIX SEVEN AND STANDING-BY
OVER

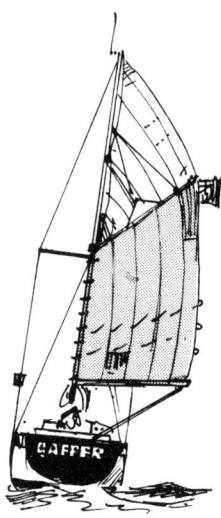

Both stations then switch to Ch. 67 (**Q**) as the yacht waits for the Coastguard to re-open communication:

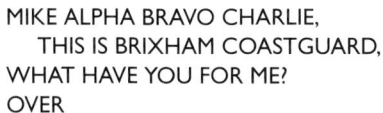

MIKE ALPHA BRAVO CHARLIE,
THIS IS BRIXHAM COASTGUARD,
WHAT HAVE YOU FOR ME?
OVER

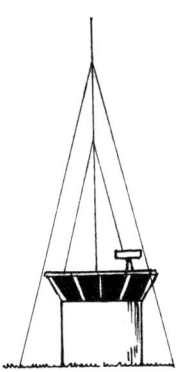

The VHF GMDSS HANDBOOK

Even a short conversation such as this looks very longwinded in print but, in practice, the whole episode would probably be over in two or three minutes.

You will notice that *every* transmission starts with the name of the station being *called*, followed immediately by the call-sign or name of the *calling* station; twice for the initial call (Q). In the case of ship-to-shore communication, the *shore station* controls communication, that is dictates the working channel (Q). In the case of ship-to-ship communication, it is the *station called* which controls communication (Q).

If no reply is received to an initial call, the calling station *must* wait three minutes before making another attempt (Q). Remember that, except in the case of distress, calls on Ch. 16 must not last longer than one minute (Q).

The VHF GMDSS HANDBOOK

Weather forecasts

In addition to the regular four-hourly (or two hourly) weather broadcasts, British Coastguards will repeat the forecast at any time upon request. However, they do become irate with people who call for a repetition of the forecast just after or before a broadcast. Since these broadcasts are interspersed with those of the Coast Radio Stations, BBC Radio 4 on 198kHz and local BBC/IBA radio stations, there is very little excuse for calling for a special, personal repetition.

Radio checks

As the transmitter section of the average marine VHF set has no in-built means of testing, most Coastguards (but not Solent Coastguard) will confirm that your transmitter is working correctly. The procedure on Ch. 16 is as follows:

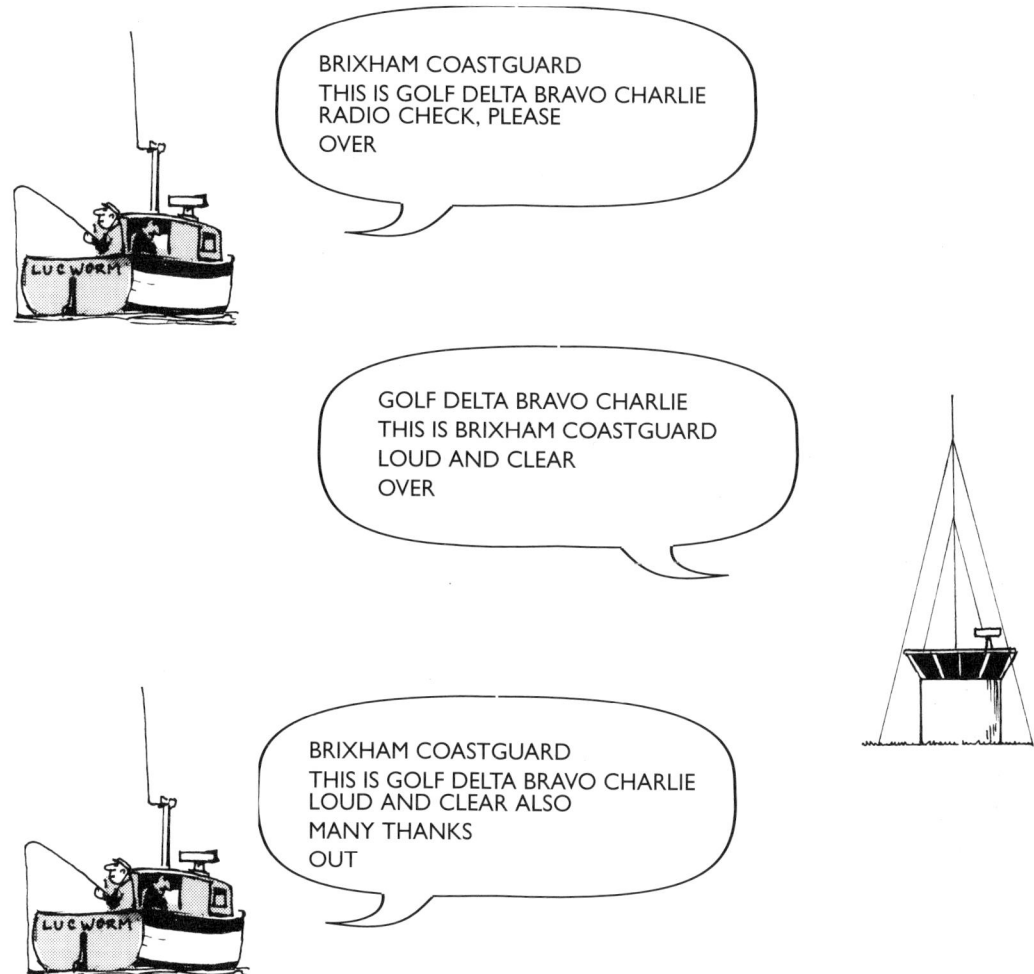

Coastguards often acknowledge this final transmission with a couple of quick presses of the microphone button without saying anything. Note that a test call (Radio Check) must not exceed 10 seconds' duration (Q).

The VHF GMDSS HANDBOOK

Local variations
Normally, the initial call to a Coastguard Station would be made on Ch. 16 (Q) but procedure is easier and quicker with Dover Coastguard who will accept initial calls directly on Ch. 69 for navigational information. Solent Coastguard accepts initial calls for routine enquiries directly on Ch. 67. However, as one person controls five radio stations between Hengistbury Head and Beachy Head, the call may have to be repeated.

Urgent calls
Any calls concerning difficulty such as dismasting, engine failure, lack of fuel, rudder problems, man overboard or diving accidents, should be addressed to a Coastguard by preceding what would otherwise have been a routine call on Ch. 16 by the phrase 'PAN-PAN, PAN-PAN, PAN-PAN' (Q).

No doubt Solent Coastguard would then ask *Gaffer* to change to Ch. 67 to ask for further details.

This type of mishap raises a question. How did *Gaffer* talk to Solent Coastguard with his aerial several feet under water at the top of his mast? The answer is that, being a prudent yachtsman, the owner would not dream of going to sea without a VHF set *and* an emergency aerial!

9

TO CALL YACHT CLUBS, MARINAS AND PORT RADIO STATIONS

Yacht clubs

In 1972, British yacht clubs and the few marinas then operating, were allocated the private channel 'M' on 157.85 MHz. With the proliferation of marinas since then, British marinas were allocated international Channel 80 in 1989 and British yacht clubs given an additional private channel 'M2' on 161.425 MHz for race and regatta control. As private channels, 'M' and 'M2' cannot be identified by international numbers, many different schemes are used. Consult the owners handbook. Channel M is often identified as '37' or 'P1' or may have a dedicated button. A variety of methods are used for selecting Ch. M2. As British yacht clubs are not licensed for Ch. 16, they *must* be called directly on Ch. M or Ch. M2 as appropriate. Try M2 first (if fitted in your set), then Ch. M if there is no reply. Address the club by name (usually 'Somewhere Base') then identify yourself by *boat's name* just like calling another boat on a working channel. For example: 'Bosham Base, Bosham Base; this is Flash, Flash. OVER'.

Marinas

All British marinas are called by name directly on working Ch. 80. This includes those on the Thames and other inland waterways. As a marina *and* commercial port, commercial ships may address Brighton Marina as 'Brighton Control' on Ch. 16 although pleasure craft must address it as 'Brighton Marina' on Ch. 80. Much information can be found in the popular yachting almanacs, of which the *Cruising Almanac* published by *Practical Boat Owner* is the most comprehensive. Most French and Spanish yacht harbours seem to operate on Ch. 09 *only*.

Port Radio Stations

Most decent, self-respecting harbourmasters throughout the World have a VHF station and welcome calls from yachts. Although there is no compulsion for yachts to ask permission to enter or leave harbour in most cases (at Dover there is!), it is only common courtesy and costs nothing.

Before calling a Port or Harbour Radio Station, however, it is vital to check all the station's details in the *Admiralty List of Radio Signals*, Vol. 6, as all ports and harbours differ.

Local variations

Most harbours are addressed as '(somewhere) Harbour Radio' but most of the major ports are addressed differently. If the station is *not* addressed as '(somewhere) Harbour Radio', the form of address will be found in ALRS, Vol.6.

Although all Coast Radio Stations work 24 hours per day, few harbours do. The hours of watch are detailed in ALRS, Vol. 6.

Calling channels differ. Some harbours/ports listen only on Ch. 16; some listen only on their working channel and some listen on both! Again, see ALRS, Vol. 6 for details. Whenever possible, the initial call should be made on the port's working channel (Q). (From 1999, initial calls to *all* shore stations will be made on their working channel.)

By far the most common port operations working channels are Ch. 12 and Ch. 14 but there are many others. Check first (ALRS, Vol. 6) to ensure that your set has the working channel(s) of the harbour/port concerned.

For example, if the yacht *Gaffer* in Chapter 8 came within the shelter of Torbay between 0900–1300, 1400–1700 and 1800–2200 (local time) any day of the week or weekend from the beginning of May until the end of September – the hours of watch of the Port Radio Station at Torquay – she could switch to Ch. 14 and call them up:

    TORQUAY HARBOUR RADIO
    TORQUAY HARBOUR RADIO
    THIS IS GAFFER, GAFFER
    CALL-SIGN MIKE, ALPHA BRAVO, CHARLIE
    OVER

If the call is to a predominantly commercial harbour or major port, it helps to identify the vessel calling as 'yacht...' in the initial radio message.

10

SHIP-TO-SHORE TELEPHONE CALLS

Coast Radio Stations act as shore-side 'telephone exchanges' to connect vessels with the world-wide telephone system or accept dictated radio-telegrams. They cannot give you harbour information (except for Jersey Radio and St Peter Port Radio, which can) and they are not particularly interested in your safety. Those throughout the UK and Northern Europe do not listen on Ch. 16 *so the initial call must be made directly on any of their free working channels* – see ALRS, Vol. 1, Part 1.

BRITISH PROCEDURE

1. Write down the telephone number you wish to call.
2. Decide on the method of payment. If you wish to use the 'YTD' system and have the cost of the call directly debited to a British telephone number, write that down too. If the call is *not* to a British telephone, via a British Coast Radio Station on VHF, a previously-arranged international Account Code must be given.
3. Find out the Coast Radio Station's working channels, in ALRS, Vol. 1 or yachting almanacs, then switch from channel to channel to find one which is available for use. This will be indicated by complete silence, that is no speech, dialling tone or 'pips' (Q).
4. Having found a free channel, *switch to high power* then press-and-hold the PTT switch for 10 seconds before releasing. All BT Coast Radio Stations are computer-controlled from one of two points in the UK. Pressing the PTT switch activates a computer which searches the UK for a free Radio Officer.
    If successful, a succession of Greenwich-like 'pips' will be heard. The 'pips' also warn others that the channel is now engaged – *for you!* (See ALRS, Vol. 1 for details of calling foreign Coast Radio Stations.)

Wait on this channel until the Coast Radio Station answers. The Coast Radio Station acknowledges by saying either:
WHO IS CALLING (somewhere) RADIO ON CHANNEL X?

or alternatively:

(Your call-sign)
THIS IS (somewhere) RADIO
WHAT HAVE YOU FOR ME?
OVER

5. With the microphone close to your mouth, press the microphone switch and *slowly* say:

(Somewhere) RADIO
THIS IS (call-sign)
MY SHIP'S NAME IS (name)
I SPELL (phonetics)
I HAVE A CALL FOR (number to be called)
MY ACCOUNT CODE IS (YTD plus Area Code and number of telephone account to be debited *or*, simply, your Accounting Authority Identification Code (AAIC))
OVER

Then release the microphone switch. The Coast Radio Station will reply saying:

(ship's name)
THIS IS (somewhere) RADIO
TRYING TO CONNECT YOU
STAND BY, PLEASE, OVER

6. Press the microphone switch and say:

(somewhere) RADIO
THIS IS (ship's name)
STANDING BY
OVER

The VHF GMDSS HANDBOOK

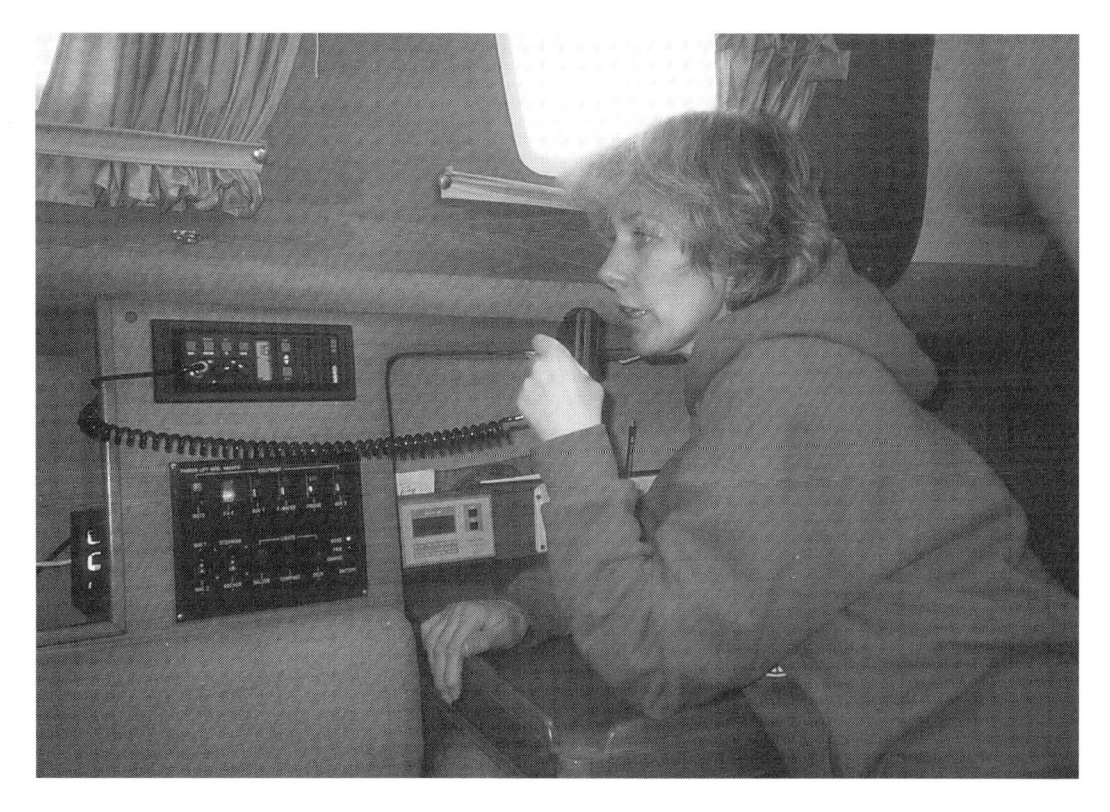

Above: When making a telephone call make sure you have all the details worked out beforehand

Then release the microphone switch. When your correspondent answers the telephone, the Coast Radio Station will say:

(ship's name)
THIS IS (somewhere) RADIO
GO AHEAD NOW, YOU ARE CONNECTED
OVER

7. Now you can press the microphone switch and speak to your correspondent. With a Simplex set, you must release the microphone switch *each time* you stop talking, as you cannot hear your correspondent as long as your microphone switch is pressed.

Your correspondent must appreciate this fact as this one-way-at-a-time conversation requires their co-operation. Be sure to warn all your possible correspondents about this beforehand as a lot of very expensive time can be wasted trying to explain the situation 'over the air'. It helps greatly if you can persuade them to say 'OVER' when they have finished talking, too, even though they do not have a switch to press.

Procedure is much easier with a Duplex set as it will automatically operate in the Duplex mode when switched to a two-frequency channel. Although the loudspeaker goes dead when the microphone switch is pressed, reception is retained in the ear piece (a telephone-type handset is always supplied with a Duplex set). Because of this, the microphone switch may be pressed and remain pressed for the duration of a normal telephone-type conversation on Duplex.

A Duplex set operates in the Simplex mode when switched to a single-frequency channel.

8. When your correspondent replaces his or her handset (only *they* can stop the call, *you cannot*), a clock in the Coast Radio Station stops recording the time which forms the basis of the charge. Do *not* immediately switch back to Ch. 16 before the CRS Radio Officer has first advised you of the duration of the call.

There is a minimum charge of three minutes.

For calls via Foreign Coast Radio Stations, the charge will be advised in French gold francs! Although French gold francs disappeared many years ago, along with gold sovereigns and guineas, this mythical currency is still used for international accountancy. Don't worry, though; your Accounting Authority will sort it all out for you (for a small fee) and send a bill in pounds sterling to the name and address which was given in the second paragraph of the Ship Radio Licence application. A table of current marine radiotelephone charges can be obtained from British Telecom (see the list of addresses at the back of the book).

> THIS IS NITON RADIO WHO IS CALLING ON CHANNEL 64?

> NITON RADIO
> THIS IS GOLF X-RAY YANKEE ZULU,
>   GOLF X-RAY YANKEE ZULU
> MY SHIP'S NAME IS *FLASH*
> I SPELL FOXTROT LIMA ALPHA SIERRA HOTEL
> I HAVE A CALL FOR 01705 123456
> MY ACCOUNT CODE IS YTD 01273 654321
> OVER

The VHF GMDSS HANDBOOK

The VHF GMDSS HANDBOOK

> **Procedure for making a VHF telephone call via a British Coast Radio Station**
>
> 1. Consult *either* the *Admiralty List of Radio Signals*, Vol. 1, Part 1 *or* a Yachting Almanac for Coast Station channels.
>
> 2. Select one of the Station's channels which is silent (no speech, 'pips' or dialling tone).
>
> 3. Press and hold PTT for 10 seconds. (On release, 'pips' indicate successful contact)
>
> 4. When the Coast Radio Station asks 'Who is calling (*somewhere*) Radio?', press PTT & say:
>
>   "......................RADIO; THIS IS ....(call-sign); ....(call-sign)
>
>   MY SHIP'S NAME IS ....; I SPELL ....(phonetics)
>
>   I HAVE A CALL FOR ....(telephone number including area code)
>
>   MY ACCOUNTING CODE IS YTD.......................(home telephone number for UK calls), *or*
>
>   GOLF BRAVO .... (numeral, numeral, for foreign calls)
>
>   OVER"
>
> 5. When the CRS says 'Go Ahead', press PTT and start speaking to your correspondent.
>
> NOTE: At end of call, stay on the Working Channel until the CRS has advised duration of the call.
>
>   Date:.......................; Time ....; Duration of Call ....minutes
>
> A number of blank forms such as this could be kept near the VHF with the call-sign, ship's name and Accounting Code already filled-in. Simply adding the other details for each call ensures fault-free procedure. The completed forms also act as a record of calls made.

Reversed charges

In addition to the 'YTD' and 'GB_ _' methods of direct payment, *transferred charge* ('reversed charge' or 'collect') calls can also be made. In this case, the recipient also pays an extra charge in addition to the cost of the call.

Radiotelephone log

Although there is no compulsion with voluntarily equipped vessels, yachts are strongly advised to keep a reasonable record of transmissions made and important messages received (Q). Officially called a 'Diary of the Radiotelephone Service', it should list:

- The operator's name with the date and times (GMT) at which he/she goes on and off watch.
- Times of departure from and arrival at ports, giving names of each.
- A summary of all communications relating to distress, urgency and safety.
- A record of communications between the vessel and shore stations or other vessels.
- Notes of important service incidents such as failures of power supply or breakdowns of apparatus.
- The position of the vessel at least once per day.

Calls to ships from the shore

To call a ship using the ordinary telephone, dial Freephone, 0800-378389. This will be answered by 'Portishead Booking' as it is, actually, Portishead Radio at Burnham-on-Sea, Somerset. Although it operates only on the short wave band, Portishead Radio co-ordinates all Britain's shore-to-ship calls

The VHF GMDSS HANDBOOK

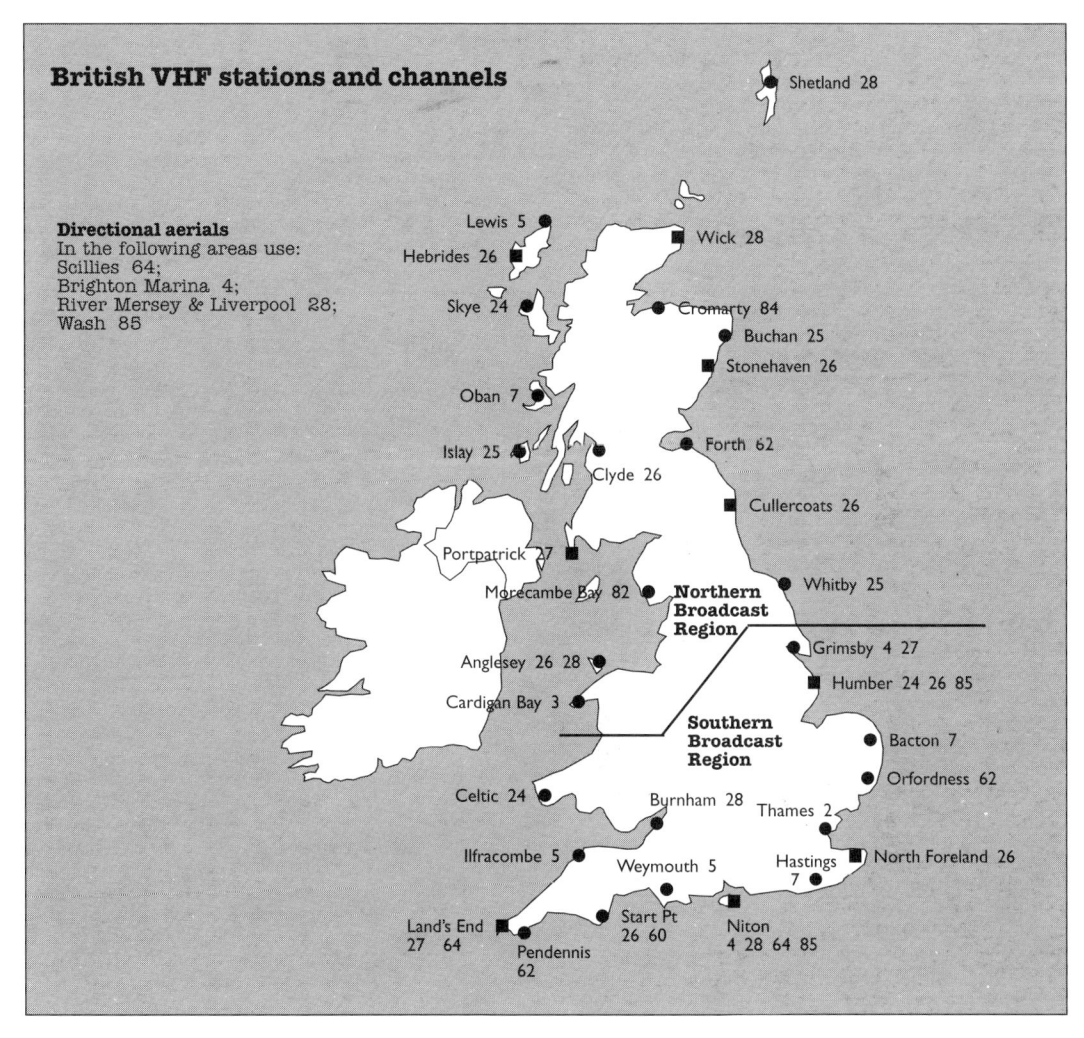

British VHF stations and channels

Directional aerials
In the following areas use:
Scillies 64;
Brighton Marina 4;
River Mersey & Liverpool 28;
Wash 85

Shetland 28
Lewis 5
Hebrides 26
Wick 28
Skye 24
Cromarty 84
Buchan 25
Stonehaven 26
Oban 7
Islay 25
Forth 62
Clyde 26
Cullercoats 26
Portpatrick 27
Morecambe Bay 82
Northern Broadcast Region
Whitby 25
Anglesey 26 28
Grimsby 4 27
Humber 24 26 85
Cardigan Bay 3
Southern Broadcast Region
Bacton 7
Orfordness 62
Celtic 24
Burnham 28
Thames 2
Ilfracombe 5
Weymouth 5
Hastings 7
North Foreland 26
Land's End 27 64
Start Pt 26 60
Niton 4 28 64 85
Pendennis 62

via Coast Radio Stations on MF and VHF as well as HF (short wave). The Radio Officer will ask for the ship's name and call-sign, approximate position or details of the voyage, name of the person called, the name and telephone number of the caller and how the call is to be charged. The caller will then be asked to 'ring off' and await re-call when the vessel has been contacted. There is a minimum charge of three minutes.

If the call is to a yacht, it is most likely that the radio will be operating in Simplex so be prepared to say 'OVER' to invite a reply after speaking!

Urgent medical advice/action

Coast Radio Stations can make free, untimed connections to doctors for medical *advice* at any time. Calls to British Coast Radio Stations on their working channels should be prefixed by PAN-PAN MEDICO spoken three times (Q). See 'Medical Advice' section of ALRS, Vol. 1 for foreign stations.

For Medical *Action* call PAN-PAN MEDICO three times, (somewhere) COASTGUARD three times, THIS IS (call-sign) three times, on Ch. 16.

11

DISTRESS SIGNAL (MAYDAY)

To qualify for a distress signal, the situation must satisfy three important conditions (Q).
- The *whole ship* must be in danger. So dismasting, heart attacks, man overboard, etc. do not qualify for MAYDAY.
- It must be *grave danger*, that is a threat to the lives of *everyone* on board. In other words a *disaster*.
- It must be *imminent danger*, that is not something which *may* happen in three or four hours' time but a disaster which is *certain* to happen during the next few minutes.

The precise circumstances of a situation can make all the difference between PAN-PAN and MAYDAY. For example; a motor yacht with engine failure ten miles south of Portland Bill, on a sunny afternoon in June in NW Force 3, is *not* in 'grave danger' although it certainly concerns the whole vessel and is certainly 'imminent' (in fact, it has just happened!). However, help may be required and the Coastguards should be alerted on Ch. 16 by preceding the call with PAN-PAN.

But the same situation, if it occurs 500 yards west of 'The Minkies' (Les Minquiers) on passage from Jersey to St Malo, in a NW Force 9, is certainly both grave *and* imminent!

Any decision to broadcast a MAYDAY signal (distress signals are always 'broadcast', that is, not addressed to anyone in particular) or transmit a PAN-PAN call must be made by the master (Q), that is the person in command of the vessel. The decision can be determined by the answer to this simple question: is my radio set about to disappear beneath the waves?

Answer yes: 'MAYDAY'
Answer no: 'PAN-PAN'

As the Mayday signal is a *broadcast* which could be received by people of all nationalities, there is a special sequence *which must be followed* if you are to be rescued quickly.
- First, the distress *call* to alert the world to your plight (Q):

> **MAYDAY, MAYDAY, MAYDAY,**
> **THIS IS (identity, identity, identity)**

For identification purposes, either the vessel's name or call-sign may be used according to circumstances. If you start to sink in crowded waters and the vessel's name is emblazoned on the side in large, clear letters then that is your best identification. However, if no other vessel is close by and you are relying entirely on radio, the call-sign is the best identification.
- Second, the distress call is ***immediately*** followed (without waiting for an acknowledgement) by the ***distress message*** (Q):

> **MAYDAY** (once again)
> **IDENTITY** (once again)
> **MY POSITION IS...** (Lat. and Long. *or* named spot *or* range and bearing *from* a well-known point of land)
> **NATURE OF DISTRESS** (e.g. 'SINKING', 'ON FIRE')
> **ASSISTANCE REQUIRED** (small boats *actually* sinking or on fire simply 'REQUEST IMMEDIATE ASSISTANCE' but, if drifting helplessly towards rocks could 'REQUEST TOW')
> **NUMBER OF PEOPLE ON BOARD** (don't distinguish between male/female, adult/child, captain/crew/passengers/kids-who-went-for-the-ride/babes-in-arms, etc. *Just count heads* including, of course, your own)
> **ANY OTHER INFORMATION** (to help people help you, you may be able to activate an EPIRB – Emergency Position Indicating Radio Beacon – or fire flares)
> **OVER** (even though you may be about to leap into a liferaft, leaving your R/T behind, the last word is always OVER!)

The VHF GMDSS HANDBOOK

To get rescued (and pass the examination) it is *absolutely vital* to give the message in the prescribed order. This is easily remembered by the following mnemonic:

- **M**: **M**ayday
- **I**: **I**dentity
- **P**: **P**osition
- **D**: Nature of **D**istress
- **A**: **A**ssistance required
- **N**: **N**umber of people on board
- **I**: Any other **I**nformation
- **O**: **O**ver

Fix this by the R/T set *and* on the back of the loo door so that it may be learnt by all on board!

Don't forget that the Mayday MESSAGE is preceded by the Mayday CALL ('Mayday' three times followed by your identity three times).

> MAYDAY, MAYDAY, MAYDAY
> THIS IS *MARY ROSE, MARY ROSE, MARY ROSE*
> MAYDAY
> *MARY ROSE*
> MY POSITION IS TWO THREE ZERO ONE MILE FROM SOUTHSEA CASTLE
> TAKING WATER AND SINKING
> REQUIRE IMMEDIATE ASSISTANCE
> 700 PEOPLE ON BOARD
> LARGE WOODEN SAILING SHIP
> OVER

12

ACKNOWLEDGEMENT OF DISTRESS SIGNALS

On hearing a distress signal (Q)

○ Listen and note the date, time and content of the message on the notepad with the pencil which is hung alongside the set – isn't it?
○ Wait 10–15 seconds to ensure that the distress signal is not acknowledged by a Coastguard or Coast Radio Station.

If it is *not* acknowledged:
○ Acknowledge receipt by saying:

> **MAYDAY**
> (Name or call-sign of distressed vessel, *three times*)
> **THIS IS** (your ship's name or call-sign *three times*)
> **RECEIVED**
> **MAYDAY**
> **OVER**

If you are able to offer practical assistance, add this information to your acknowledgement.
○ Tell the world, by saying on Ch. 16:

> **MAYDAY RELAY, MAYDAY RELAY, MAYDAY RELAY**
> **THIS IS** (identity), (identity), (identity)
> **THE FOLLOWING DISTRESS MESSAGE WAS RECEIVED FROM** (distressed vessel) **AT** (time)
> **MESSAGE BEGINS**..............................
> **MESSAGE ENDS**
> **OVER**

If you hear the distress signal acknowledged by a Coastguard or Coast Radio Station and are able to offer assistance *call them* by preceding their name by the word MAYDAY. (All distress messages are preceded by the word MAYDAY).

If you hear the distress call acknowledged by a Coastguard or Coast Radio Station and are *not* able to render assistance, *keep quiet*.

On seeing a distress signal

Broadcast the information by saying on Ch. 16:

> **MAYDAY RELAY, MAYDAY RELAY, MAYDAY RELAY**
> **THIS IS** (your ship's name or call-sign *three times*)
> **MY POSITION IS** (Lat. and Long. *or* range and bearing *from* a well-known point of land)
> (Type of distress signal seen)
> (Time distress signal was seen)
> (Position of distress signal seen or bearing *from* your own position)
> **OVER**

13

THE GMDSS – AN INTRODUCTION

To appreciate the need for the radical change that is about to hit the world of marine radio, it is useful to look back 100 years.

From 1898 to 1914, radio was a newfangled gimmick only fitted to ships – mainly large passenger liners – on a voluntary basis principally for the amusement of wealthy passengers to send and receive radiotelegrams by Morse Code (there was no R/T until the 1920s) and to catch Dr Crippen in 1910. Safety of life at sea was very much a secondary consideration. In any case, marine search and rescue was a very disorganised and haphazard affair. The crews from a number of small ships had been rescued as the result of a radio message but it was the *Titanic* disaster of 1912 which dramatically illustrated the enormous benefit of radio at sea. Disaster though it was (only 705 lives were saved out of 2,227), it would have been an even bigger disaster had it not been for the Marconi Wireless Operator, Jack Phillips, who perished with most of the crew, tapping out the very first SOS. (Previously, the Distress Signal had been CQD but it had just changed so Mr Phillips sent both signals to make sure!)

THE FIRST SAFETY OF LIFE AT SEA CONVENTION

The World was devastated by the *Titanic* disaster. It triggered the first international *Safety of Life at Sea Convention* (SOLAS) which was signed in January, 1914. One of the many regulations introduced in the Convention was to make radio compulsory on all passenger ships over 1,600 gross registered tons (grt) on international voyages. With the development of radiotelephony in the late 1920s, the lower limit was (and still is) set at 300 grt (Q) to include cargo ships and **all** passenger ships (now defined as carying more than 12 passengers) on international voyages. Small craft, such as yachts, below these limits and offshore installations which carry radio on a voluntary basis, are known alternatively as non-SOLAS, non-Convention or as voluntarily-equipped.

Please Note: Marine radios are required for small British commercial vessels (including sail training yachts) under the Department of Transport's Maritime and Coastguard Agency (MCA) *Code of Practice*.

THE CONCEPT OF THE GMDSS

Although marine radio has been instrumental in the rescue of many thousands of lives at sea, it was more by good luck than good management. Until recently, there was no co-ordinated world-wide search and rescue organisation; countries did their own thing, if they did anything at all! In 1979, the International Conference on Maritime Search and Rescue invited the International Maritime Organisation (IMO) to develop a worldwide marine search and rescue scheme. Thus was born the Global Maritime Distress and Safety System (GMDSS). Introduced on 1st February, 1992, it is due to be fully implemented on 1st February, 1999. At the time of writing (February 1998), we are in the transitional period between two quite different sets of SOLAS Regulations. Like most things which seemed a good idea at the time, there are sure to be some fine tuning changes as the GMDSS settles down in the 21st Century. So do watch the yachting press and send your SAE to Fernhurst Books for updates!

Deadline
With only a few months to go however, there is increasing doubt that the deadline of 1st February, 1999, may be met due to a huge backlog of ships

which must be re-equipped with new GMDSS equipment and deck officers who must be trained as radio operators. Under the present GMDSS timetable, the use of Morse code *at sea* will cease officially at midnight UTC on 31st January, 1999. From then on, marine radio communication will only be either by voice or by Telex and the equipment will be operated by the deck officers on the bridge. (The profession of ship's Radio Officer will then pass into history along with stokers and trimmers, etc.) **Listening watch will cease**. Watch on the Calling and Distress channels/frequencies will then be kept electronically by a dedicated watchkeeping receiver. On hearing an audible alarm, a deck officer will drop his/her current task and assume the duties of radio operator. Merchant ships are simply following the lead set by airliners 30 years ago.

Sea Areas

Since 1914, merchant ships have had to be fitted with radio equipment appropriate to their *size*. The minimum range was (and until 1999 still is) a mere 150 miles! From 1st February, 1999, however, merchant ships must be able to contact the shore **from any part of the World by at least two independent means**. Under GMDSS, all ships over 300 grt (Q) and all passenger-carrying vessels on international voyages must carry radio equipment **appropriate to where they sail** no matter what their size. Accordingly, the World is divided into four Sea Areas:

Sea Area A1 is defined as *within radiotelephone range of at least one* VHF *shore station with DSC facilities* (Q). As VHF range is dependent on the heights of the respective aerials (see Chapter 5), no general range can be given. Each station declares its own range. Some quote 20 miles, but a few quote 100 miles. The average range *to a merchant ship* is about 35 miles. Full details of each station are given in the *Admiralty List of Radio Signals* (ALRS), Volume 5 (Q). Obtainable from Admiralty Chart Agents and nautical bookshops, it also contains an excellent description of the GMDSS and is well worth the current (1998) price of £23.

Sea Areas A2, A3 and A4 cover *offshore, oceanic and the polar regions* respectively. These are outside the scope of this book but are covered by the companion books *Marine* SSB *Operation* (GMDSS edition) by the same author (ISBN 1–898660–40–9) and GMDSS *for Small Craft* by Alan Clemmetsen (ISBN 1–898660–38–7), both of which are published by Fernhurst Books.

NEW TECHNOLOGY

During the past 100 years, marine radio has always taken advantage of the latest technology. With the introduction of the GMDSS, the opportunity was taken to incorporate the latest digital technology into **calling** and **Distress Alerting** techniques and frequencies. It must be emphasised that the new technology is **not a communication system**. Traditional voice **communication** procedure, frequencies and channels are not being changed. The new technology is called **Digital Selective Calling** (DSC) which requires new equipment. It may be built-in to the VHF set or connected as a separate DSC Controller. Radio engineers have long appreciated that digital technology is far superior to analogue technology by being more efficient, effective and reliable. We all now appreciate how much better Compact Discs are than vinyl LPs, and the same may be said of GSM mobile telephones when compared with analogue mobiles. One of the reasons why Morse code survived for so long was that (in the, as some say, good old days) the digits, (the dots and dashes) were tapped-out slowly by the hand of a highly-skilled specialist with three years training. Now the digits can be generated automatically 1,000 times faster at the touch of a button by anyone with just a few minutes training.

DSC Controllers

Compulsory-fit merchant ships
For compulsory-fit merchant ships over 300grt, DSC Controllers fall into two classes:

Class A   Class A Controllers can control MF, HF and VHF radios; and
Class B   which can control MF and/or VHF radios on ships not required to fit a Class A Controller.

Small Craft
So where does all this leave us 'Yotties'? Being (usually) under 300 grt, until very recently the short answer was, officially, nowhere! It must be

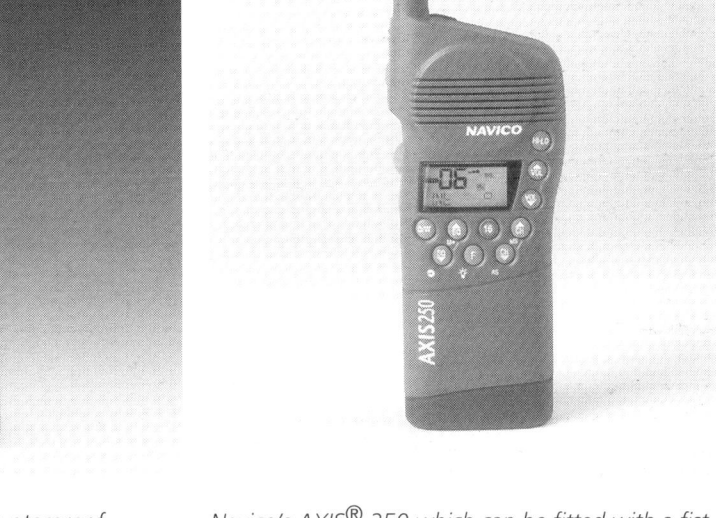

The McMurdo R1 VHF GMDSS waterproof transportable for on-scene SAR communication

Navico's AXIS® 250 which can be fitted with a fist mike, is a full 55 channel R/T

appreciated that the GMDSS was devised by merchant seamen *for* merchant seamen who simply do not recognize the existence of small craft. Consequently, **no provision was originally made for small craft under 300 grt!** If we wished to 'join' the GMDSS, the only equipment available was the very expensive kit used by the professionals. Although outline plans and the timetable for the GMDSS were published in 1987, no yachting organisation seemed interested in translating GMDSS requirements into small craft terms. In January 1995, Mr Bill Sandford, the then Chairman of the Small Craft Group of the Royal Institute of Navigation, became very concerned at the lack of interest by other bodies and formed a committee of yachting interests to draw up a desired specification for non–SOLAS VHF GMDSS equipment. The resulting UK Performance Standard MPT 1279 for what is called 'Class D' equipment was not published until mid-1996. This was subsequently developed into the European Specification EN301025. Only then could radio manufacturers start to design reasonably priced DSC VHF sets for the leisure market. It then takes at least two years to develop, approve and produce radios for sale. The facilities of Class D Controllers are only slightly less than those of Class B Controllers. Note: Fixed, Class F sets are not approved for use in European vessels. They are available outside Europe (notably the USA) but should not be bought due to their very poor specification.

Transportable Sets

Approved hand-held portable sets will be made to the Class F Standard. This gives very limited facilities for making DSC Distress Alerts (MAYDAY), All Ships Urgency (PAN-PAN) and All Ships Safety (Sécurité) calls. Their MMSI will be coded to indicate that it is a portable set which may be used on a number of different vessels. However, they will not have the capability of automatic repetition of Distress Alerts or of automatic position and time entry into Distress Alerts, Nor will they be able to receive DSC calls or DSC Acknowledgements.

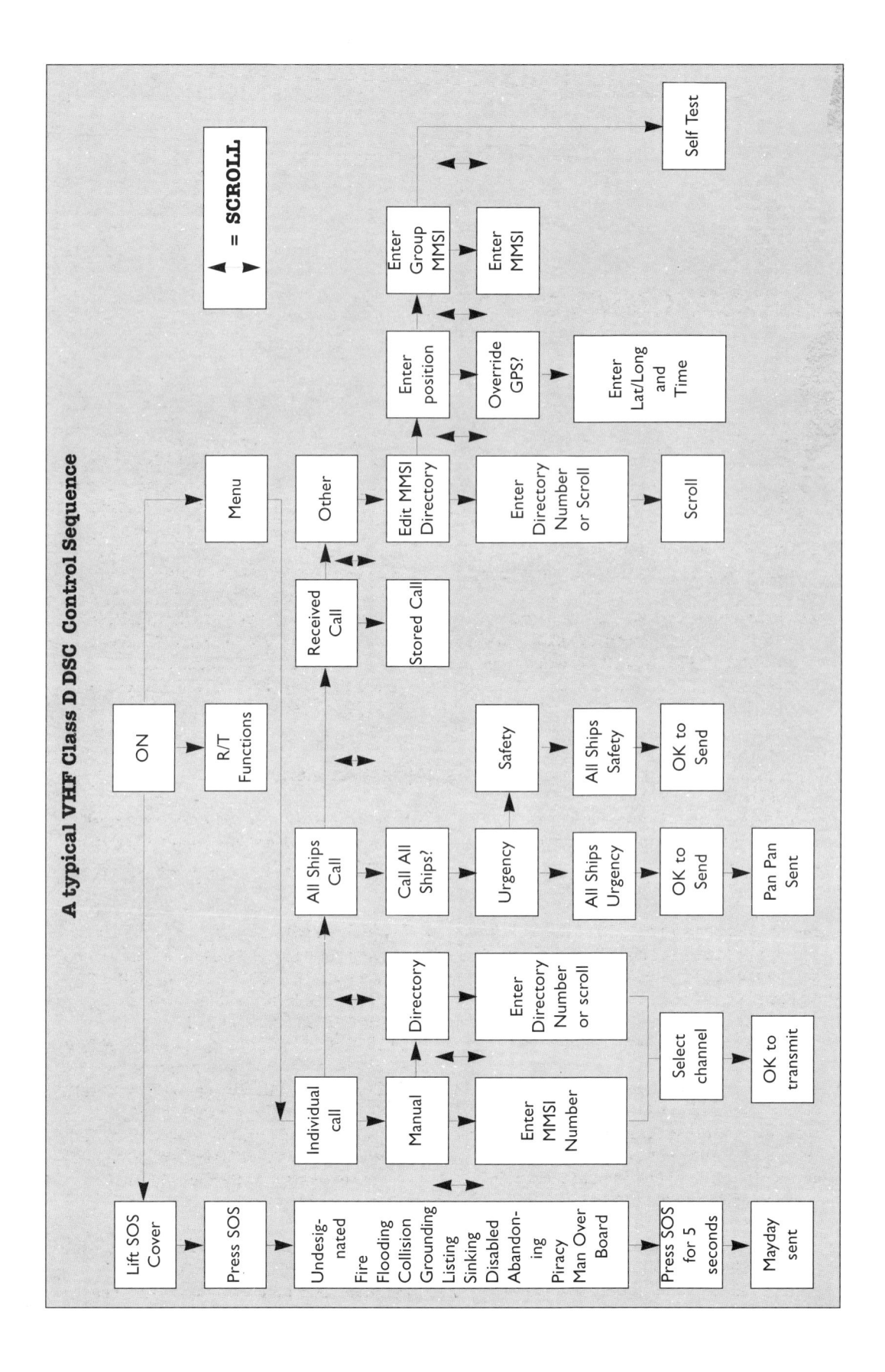

A typical VHF Class D DSC Control Sequence

◄―► = SCROLL

14

THE DSC CONTROLLER

Different makers adopt different layouts but they all conform to the requirements of MPT 1279. There are several new features on all the new sets and Controllers:

1. A large screen displaying a lot of information about the current status of the set such as:
   a. The present position and time if the set or DSC Controller is integrated with a GPS receiver or a recent manually-entered position and time. **These are automatically included with the new DSC Distress Alert.**
   b. The channel currently selected.
   c. The type of an incoming call, the MMSI of the calling station and the channel for subsequent voice communication. (Class D Controllers incorporate a Channel 70 watchkeeping receiver.)
   d. The vessel's own MMSI.
   e. Whether the set is receiving or transmitting.
   f. The transmitter output power selected.
2. A 12–button numerical keypad. This is used for selecting channels and transmit power, entering the MMSI of a station to be called, and entering the current position and time if not interfaced with GPS (Q).
3. A large red button marked **DISTRESS** or **SOS** guarded by a spring-loaded cover. When first pressed, a menu of ten different distress situations is opened.
4. Buttons marked CALL (or MENU), OK (or ACCEPT or ENTER), CANCEL and two arrow buttons for scrolling up and down through the Menu (or left and right).
   a. The CALL/MENU button opens the main menu of four options which are selected by the arrow buttons. The options are:
      i. Individual Call (the default)
      ii. All Ships Call (for All Ships URGENCY/SAFETY)
      iii. 'Log' for received calls, and
      iv. 'Other' for what are called 'housekeeping' functions. From these four main options, further sub-menu options can also be selected with the aid of the arrow buttons.
   b. The OK/ACCEPT/ENTER button accepts the displayed menu item.
   c. The CANCEL button cancels all selected DSC functions and reverts to the initial display. All selected functions are automatically cancelled if not accepted within 5 minutes.

Basic Functions

As the course of instruction includes practical operation of Class B (for ROC) and Class D (for SRC) Controllers, only outline details are given here.

1 R/T Operation

Upon switching-on (usually by means of the combined on/off and volume control), full R/T functions are available together with the DISTRESS/SOS and CALL/MENU buttons. Channels and transmitter output power (1W/25W) are selected by pressing the appropriate buttons on the keypad. A DUAL Watch (DW) button is also available, as is a squelch control. The TRANSMIT/RECEIVE function is selected by the usual PRESS–TO–TALK (PTT) button on the side of the fist microphone or in the handle of a telephone-type handset.

2 DSC Distress Alert (MAYDAY)

a. Lift and hold the cover over the big red DISTRESS/SOS button with one hand then

*ICS Electronics' **DSC3** Class D Controller which operates in conjunction with their **VHF3** VHF set*

quickly press–and–release the red button with the other hand. So as to reduce the possibility of false Alerts, pressing the DISTRESS/SOS button once does not send a DISTRESS ALERT. Pressing the red button once merely opens a menu of 10 disasters plus an 'undesignated', i.e., an unspecified, option. By default UNDESIGNATED always appears first. If time allows, the other ten items on the menu are revealed one at a time by means of the two SCROLL up/down buttons on the panel.

b With an appropriate disaster displayed, press the red button a second time and hold depressed for 5 seconds to send a DISTRESS ALERT. The vessel's MMSI, its latest position, the time at which that position was entered and the nature of the distress (from the menu) is then transmitted 5 times automatically on Channel 70 (Q). In this way, the minimum information required to initiate a Search–And–Rescue (SAR) operation is transmitted at the touch of a button.

The VHF GMDSS HANDBOOK

Having sent the DISTRESS ALERT, the VHF set then automatically selects Channel 16 for a back-up traditional MAYDAY broadcast, if time allows. This is to supply further information such as the number of people on board, the action being taken and so on, and for subsequent on-scene communication.

NOTE: Channel 70 is for Digital Selective Calling ONLY, it CANNOT be used for R/T.

Normally, the DSC DISTRESS ALERT will be manually acknowledged immediately (also by DSC) by a Maritime Rescue Co-ordination Centre (MRCC) (Q) or Maritime Rescue Sub-Centre (MRSC) (Q). On VHF these will be Coastguard Stations. If a DSC Acknowledgement is not received after about 4 minutes, the DISTRESS ALERT is automatically repeated every 4 minutes until it is either acknowledged or the vessel sinks (Q). The DSC Acknowledgement is displayed on the receiving set's screen and an alarm sounds. (In exceptional circumstances – such as being out-of-range of the shore – a DSC DISTRESS ALERT may be acknowledged by a merchant ship. This can only be done after a second DISTRESS ALERT attempt has not been acknowledged by a shore station (Q)).

Note 1: Class D Controllers do not have DSC DISTRESS ACKNOWLEDGEMENT or DSC DISTRESS RELAY facilities.

Note 2: If a false DISTRESS ALERT is inadvertently sent, IMMEDIATELY SWITCH THE SET OFF TO PREVENT ANY FURTHER REPETITIONS. Then switch the set on again, select Channel 16 and cancel the ALERT immediately by voice. The call, preceeded by the word MAYDAY (once) may be addressed to 'All Stations' (three times) or a particular MRCC/MRSC (Q).

3 Routine DSC Calls

Routine DSC calls may only be made to an individual station – either ship or shore. The Address MMSI must be entered manually with the keypad or selected from an internal operator-entered directory of frequently-called ship or shore station MMSIs. (The international directory is the ITU *List of Ship Stations*. In three volumes from Admiralty Chart Agents, it lists all the World's radio-equipped vessels by name in alphabetical order together with their call-signs and MMSIs.) The DSC Acknowledgement from the station called indicates a suitable channel for subsequent *voice communication to which the receiving set automatically tunes.*

4 'All Ships' Urgency and Safety (Pan-Pan & Sécurité)

This is the second option on the main Menu. By default, ALL SHIPS URGENCY (PAN–PAN) is the first option on the sub-menu. ALL SHIPS SAFETY (SÉCURITÉ) is selected by the arrow buttons.

5 'Log' or 'Received Call'

A number of DSC calls received on Channel 70 are automatically recorded in an electronic log. For review, this is the third option on the main Menu as selected by the arrow buttons. If a DSC DISTRESS ALERT is received, listen on Channel 16 for the subsequent voice MAYDAY and voice acknowledgement from an MRCC/MRSC (Q). If nothing is heard after about 15 seconds, call the distressed vessel **by voice on Channel 16** to acknowledge receipt and offer assistance (Q). Then broadcast a MAYDAY RELAY by voice on Channel 16 to ALL SHIPS or a particular MRCC/MRSC (Q). If a voice acknowledgement is heard from an MRCC/MRSC, **call them by voice on Channel 16** to offer assistance if able (Q). If you aren't able, KEEP QUIET! Remember all Distress messages start with the word MAYDAY once.

6 'Other' Options

This fourth option in the main Menu is used for 'housekeeping' functions such as entering an MMSI into the Directory, manually entering a position and time in the absence of a GPS, entering a Group MMSI and performing an internal self-test routing (Q).

SIMULATOR

For pre-course familiarisation and subsequent practice, a low-cost computer simulation program of a Class D controller is available on floppy disc from LightMaster Software on 0181–653–0218; E–mail: martinq@dircon.co.uk The PC requirements are very modest.

Restricted Operator's Certificate (ROC)

The written paper for the ROC is very similar to the SRC but the practical test must be conducted on a GMDSS Class A or Class B Controller. These have facilities for sending a DSC Distress Acknowledgement and a DSC Distress Relay in addition to the facilities of a Class D Controller.

15

NAVTEX, EPIRBs and SARTs

NAVTEX

This is a system for broadcasting *free* **M**aritime **S**afety **I**nformation (MSI) such as gale warnings, navigation warnings, weather forecasts, etc. in printed form. The name can be thought of as standing for **NAV**igation **TEX**t. The broadcasts, on 518 kHz (Q), are made by many (but not all) of the World's medium-wave Coast Radio Stations. The UK is covered by 3 stations at Niton–S (Isle of Wight), Cullercoats–G (Tynemouth) and Portpatrick–O (near Stranraer in SW Scotland). As well as these, stations at Corsen–A (Brittany), Ostend–T (Belgium), Ijmuiden–P (Netherlands) and Rogaland–L (Norway) also serve the English Channel, the Irish Sea and the North Sea. The daytime range is about 300 miles (Q) but can be much greater at night.

Obviously, they cannot all broadcast on the same frequency at the same time. So, they operate on a 'time-share' basis with each station enjoying an exclusive 10 minute slot in every 4 hour period. Even so, there are still insufficient slots for every station in the World so there are many simultaneous transmissions. To prevent mutual interference, all stations sharing the same slot are spaced well away from the others. For example, stations in Peru and China share the same slot as Niton Radio. Fortunately for us Brits, the language used on NAVTEX print-outs is English (Q); in fact, English is the only language authorised for use in the whole of the GMDSS.

Although the 518 kHz transmissions can be received on some SSB receivers and displayed on a Personal Computer using special software, it is strongly recommended that a dedicated receiver/printer with its own special aerial (which can be mounted on the pushpit) is fitted. They consume so little current (0.3A) that they can be left switched on all the time at sea. Their cost is surprisingly inexpensive compared with the benefits of, say,

NAVTEX *Codes*
A Coastal Navigation Warnings
B Meteorological Warnings
    (e.g., Gale Warnings)
C Ice Reports
D Distress/Search and Rescue Reports
E Meteorological Forecasts
F Pilot Service Messages
G DECCA Messages
H LORAN–C Messages
I OMEGA Messages
J SATNAV Messages (GPS)
K Other Electronic Navigation Aid Messages
L Area Navigation Warnings
    (in the UK, Drilling Rig Movements)
V Amplification of Navigation Warnings
    Initially Announced under '**A**'
W Special Services
X Special Services
Y Special Services
Z No Messages on Hand

having a print-out of the Shipping Forecast from a small 'till roll' twice daily and quite automatically!

Every message is identified by a four-character group (Q). The first character, a letter, identifies the station (Q). For example, **S**A32 identifies the transmitting station at **Niton Radio**. The second letter indicates the type of message (Q), for example, **A** means **Coastal Navigation Warnings**. The figures are merely the serial number of the message (01–99). The same message number will be ignored for the next three days to save paper, *provided that the receiver remains powered*. The next time that number appears, it will probably be a different message. Messages numbered **00** signify urgent messages such as Gale Warnings and SAR Messages which are printed on receipt as well as at their scheduled time (Q). This causes an alarm to sound.

The VHF GMDSS HANDBOOK

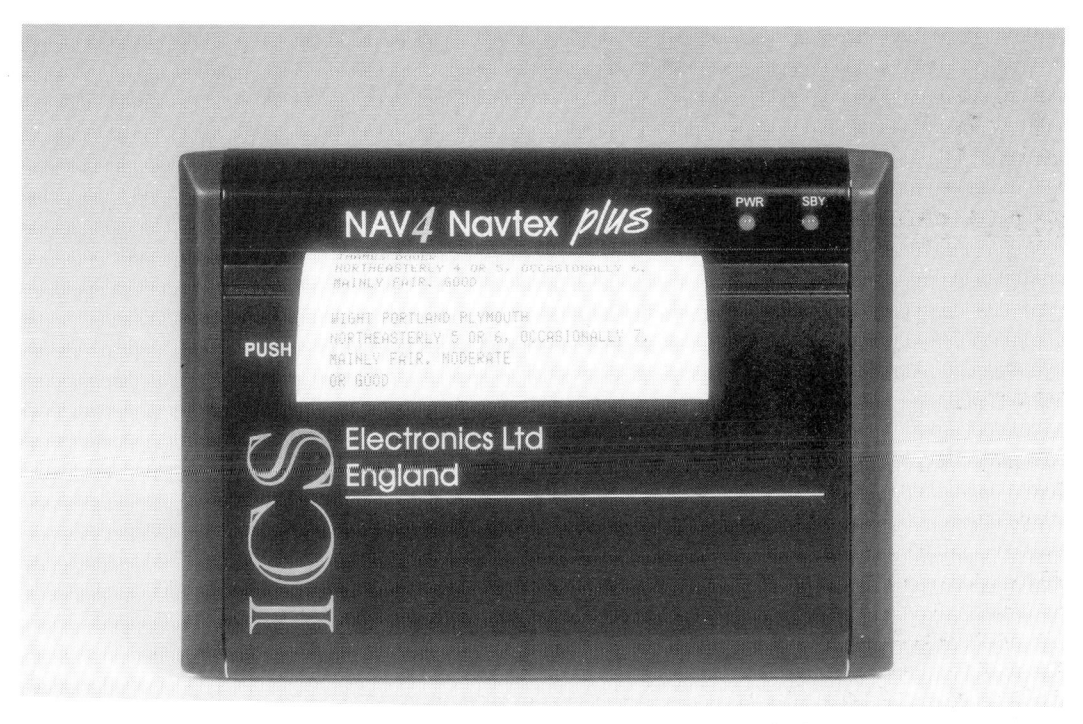

A typical small craft NAVTEX receiver

```
NAVTEX MESSAGE = = = = = = = = = = AD97
010357 UTC JUN 97
THIS IS MRCC ETEL (FRANCE)
        PHONE  :  (33) 2 97 55 35 35
        TELEX  :  (42) 950519
        FAX    :  (33) 2 97 55 49 34

EPIRB 121.5 MHZ REPORTED IN POSITION:
LAT 47 12N, LONG 004 57W
SHIPS IN THE VICINITY ARE REQUESTED TO
HAVE A SHARP LOOKOUT AND REPORT
ALL INFORMATION TO MRCC ETEL
+


NAVTEX MESSAGE = = = = = = = = = = AD98
011426 UTC JUN 97
MAYDAY MESSAGE HAS BEEN SENT BY:
BARGE JOLI
IN DISTRESS IN POSITION 48 02N - 007 30W
3 POB
ALL SHIPS ARE REQUESTED TO PROCEED TO THIS
AREA AND TO REPORT ALL INFORMATION TO CORSEN
MRCC BY VHF CHANNEL 16 OR THROUGH BREST LE
CONQUET RADIO ON HF 2182 KHZ
+
```

```
NAVTEX MESSAGE = = = = = = = = = = SA72
WZ 595
DOVER STRAIT  RIDENS SOUTH EAST
LIGHTBUOY 50-43N 01-19E OFF
STATION +

NAVTEX MESSAGE = = = = = = = = = = SA78
WZ 631.
FRANCE, NORTH COAST. CHERBOURG,
NORTHWARD. DANGEROUS WRECK REPORTED
49-48.0N  01-26.0W. UNAUTHORISED
NAVIGATION PROHIBITED WITHIN
1.5 MILES OF THIS POSITION. +

NAVTEX MESSAGE = = = = = = = = = = TA55
290920 UTC JUN =
OSTENDERADIO INFO 154/95 =
RADIOBEACON NORTHEAST OF THE LIGHTHOUSE OF
OSTENDE IS UNRELIABLE +

NAVTEX MESSAGE = = = = = = = = = = GA83
WZ 1509
DOVER STRAIT
RUYTINGEN SOUTHWEST BUOY 51-05N 01-47E
DAMAGED AND UNLIT
```

An actual (if a little dramatic!) sequence of NAVTEX messages

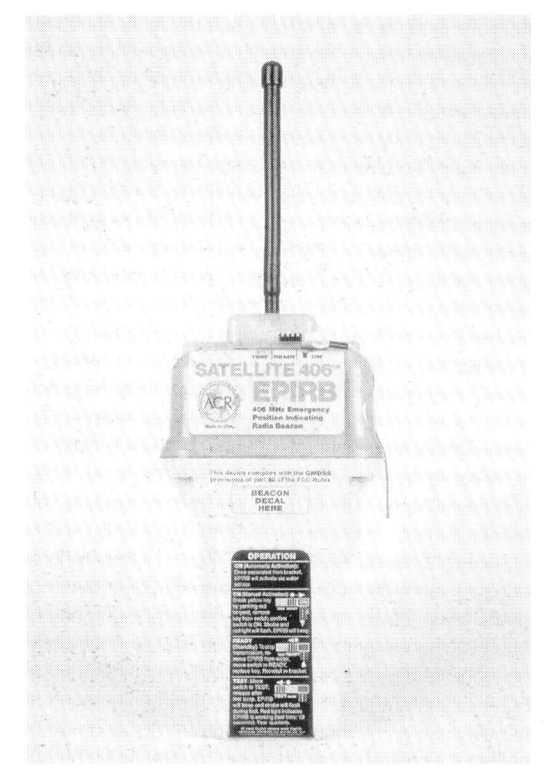

EPIRB

There are several types of **E**mergency **P**osition-**I**ndicating **R**adio **B**eacon. The main purpose of these small, buoyant devices is to transmit an automatic Distress Alert and indicate the position of survivors (Q). The two of most interest to small craft transmit on 121.5 MHz and 243 MHz (the aircraft Distress frequencies) and 406 MHz (Q).

121.5 MHz EPIRBs

These are about the size of mobile phones and are primarily intended for locating crashed aircraft *by* rescue aircraft which can home-in on the signal. Aviators know them as **E**mergency **L**ocation **T**ransmitters (ELTs) but they are also used by fell-walkers and mountaineers who call them **P**ersonal **L**ocator **B**eacons (PLBs)! Their signals are not, normally, picked-up by aeronautical ground stations. Civil airliners do not, usually, maintain a listening watch on 121.5 MHz until well beyond Air Traffic Control range. This means 150 miles west of

Below: ACR Electronics Inc's Mini B$_2$™ EPIRB transmits simultaneously on 121.5 MHz and 243.0 MHz is seen here with a C–Light™

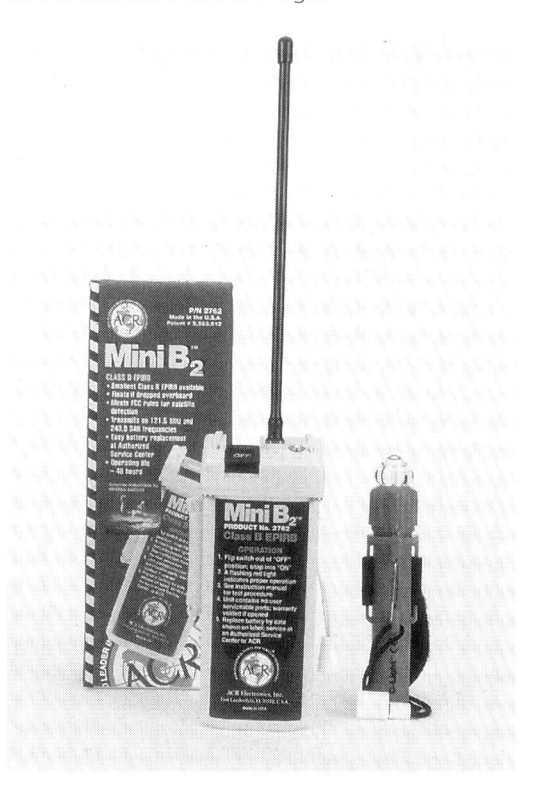

Above: ACR Electronics Inc's Satellite 406 MHz EPIRB

When a new NAVTEX set is purchased, it is programmed to receive all stations and print all messages. At any time, all stations of no interest and some irrelevant message types can be deselected. Message types are listed above. Those coded **A**, **B**, **D**, and **L** cannot be deselected. Station names, positions, identification letters, range and service areas are published in the *Admiralty List of Radio Signals*, Vol 5, but their transmission schedules are listed in ALRS, Vol 3, under 518 F1B.

Corrupted Reception

By being transmitted twice, NAVTEX messages are extremely reliable. Occasionally, however, a few characters may be recieved incorrectly. In this case, an asterisk is printed instead of the wrong character. If more than 5% of the message is corrupted, the message is not printed.

Testing

A print-out of an Internal Test Report can be obtained by following simple instructions (Q).

Tony Bullimore, the Vendée Globe yachtsman, attributes his rescue by the Royal Australian Navy in the Southern Ocean to the Satellite 406™ EPIRB he used

Shannon or Prestwick! Even if a civil airliner did pick-up one of these signals, at 30,000 feet it cannot locate the position closer than several thousand square miles! The signals can also be picked-up by one of four SAR COSPAS/SARSAT satellites in near-Polar orbit. The satellite then relays the signal to a ground station. The EPIRB can also be located to within 15 miles, although this takes 3 hours. However, there are vast areas of the North Atlantic and Pacific Oceans and most of the Southern Hemisphere without coverage due to their low orbit. As Distress alerting devices for small craft, they are therefore pretty useless! Having broadcast a Distress Alert by DSC or R/T, however, the SAR aircraft or offshore RNLI lifeboat can quickly 'home-in' on the 121.5 MHz signal. So, although poor for Distress alerting, these 121.5 MHz EPIRBs are superb for pin-point location!

406 MHz EPIRBs

These latest GMDSS-approved devices are much larger (and more expensive) than the 121.5/243 MHz type but a different animal altogether with many advantages. Their signals are **only** picked up by the COSPAS/SARSAT satellites and relayed to an appropriate ground station (Q). But, by operating in a store-and-forward mode, the whole World is (eventually) covered. These satellites also fix the position of the EPIRB to within approximately 3 miles. This takes between 30 minutes and two hours. Only then can the worldwide Search and Rescue System be initiated and that can take another hour! So do not hold your breath whilst awaiting the Sea King or Nimrod after throwing the switch.

Another big advantage over the 121.5/243 MHz type is that this type identifies the vessel. Some countries do this by programming-in the vessel's MMSI but the UK and USA do it by including the EPIRB's serial number with the information transmitted. When purchased new, a supplied registration card is completed with details of the vessel and its owner for posting to a national SAR Registry. In the UK this is Falmouth Coastguard. **If a used 406 MHz**

EPIRB is obtained, it is vital that the Registry is advised of the change of ownership and vessel. Otherwise the SAR aircraft could be looking for a supertanker instead of a small yacht!

As an optional extra, 406 MHz EPIRBs **usually** also emit a 121.5 MHz signal for pin-point location. It is important to check for this facility for when buying. For merchant ships, float-free housings are available but **these are not recommended for small craft**. Small craft are advised to keep the EPIRB in the 'panic bag' for carrying into the liferaft. There it should be manually activated, secured to the liferaft with the cord provided **and only then placed in the water** where it will float close by the liferaft. It should not be used inside the liferaft as the wet canopy shields both the signal *and* the brilliant flashing white strobe on the top of the casing. Some types incorporate a water-activated battery – obviously, great care must be taken to store them in a dry place.

Testing

All 406 MHz EPIRBs should be tested monthly (Q) *without making a live transmission* using the means provided. Read the instructions on the EPIRB. This usually involves operating a spring-loaded switch and observing a light (Q). The battery expiry date should also be checked at the same time (Q).

Transport

For carriage to or from the boat, the battery should be removed to prevent accidental activation. If this is not possible, aluminium kitchen foil should be doubly wrapped around the aerial to prevent accidental radiation. In the event of accidental activation, the nearest MRCC/MRSC should be informed immediately **before switching-off** (Q).

SART

These devices should really be called SARRTs because they are **S**earch **A**nd **R**escue (**RADAR**) **T**ransponders (Q). In the GMDSS, they are the main means of locating distressed vessels or their survival craft (Q). About the same size as the 406 MHz EPIRB (say, 350mm), these devices house a 9 GHz or 3 cm receiver/transmitter. On receiving a 9GHz radar signal, the transmitter is triggered to return a coded signal to the interrogating radar (Q). The returned, enhanced signal paints a distinctive line of 12 dots on the radar screen of the approaching ship or aircraft (Q). These dots

McMurdo's E3 406 MHz EPIRB

radiate from the SART's position outwards to the scale around the edge of the screen to indicate the bearing (Q). At about 1 mile from the SART, the outer end of the line of dots opens to form a fan shape. As the separation decreases further, the 'fan' widens to form 12 concentric circles centred on the approaching ship's position (Q). At a minimum height of 1m asl (for which a small mast is provided) their range is about 5 miles to merchant ships and about 30 miles to searching aircraft at 3,000 ft (Q).

In the 'idling' condition, a white strobe flashes once every 2 seconds. When the transmitter is activated by an interrogating radar, the light becomes steady and an internal beep sounds. This must be very reassuring to survivors in a liferaft!

Like the EPIRB, it must be deployed **outside** the canopy of a liferaft (for which a hole should be provided) and it should be secured with the cord provided. A SART must not be used in conjunction with a radar reflector as this could mask the characteristic 12 dots on the radar screen or shield the SART from the incoming signal (Q).

The VHF GMDSS HANDBOOK

Liferaft with a McMurdo SART deployed

Testing

SARTs are also a mandatory part of a ship's GMDSS equipment and should be tested monthly without breaking the seal covering the switch (Q). Read the maker's instructions to activate it for test, then hold the SART aloft in view of your own 9 GHz radar. Ask another person on board to check for the characteristic 12 concentric rings on the screen. When successful, switch off by following the manufacturer's instructions (Q). At this point, take the opportunity of checking the expiry date of the battery (Q). The battery should be capable of operating the SART for 96 hours in the 'idle' condition followed by a minimum of 8 hours continuous transmission once activated (Q).

HAND-HELD VHF

EPIRBs and SARTs taken into a liferaft considerably increase the chances of survival and a hand-held VHF for on-scene communication is the 'icing on the cake'. Several waterproof hand-held VHF transceivers are now on the market and could prove invaluable in a distress situation. However, beware of the GMDSS-approved models. Being made specifically for SAR operations, they only have a limited number of single-frequency channels. A waterproof version of a non-GMDSS hand-held VHF (or even a standard hand-held VHF in an inexpensive waterproof bag) could double-up as a second set for use in the cockpit, flying bridge, dinghy or liferaft and as a stand-by in case of dismasting.

MOBILE TELEPHONES

With their reduced size, weight and cost, the use of mobile telephones has vastly increased in recent years and inshore sailors may be tempted to take them to sea in place of a proper marine VHF set. Although not intended for use at sea, their service area does 'spill over' a little onto low-lying coasts where they are ideal for public telephone calls. However, mobile 'phones are unsuitable for safety purposes as they suffer many serious disadvantages when compared with a proper marine VHF set.

Some of the disadvantages of using mobile phones at sea

Calls can only be made to specific numbers and they need your number to call back. Broadcasts cannot be made to other craft in the area. You may be able to call a Coastguard directly (chargeable!) or through the 999 system but fellow yachtsmen in your area, who may be able to help, will not hear your plea. **You cannot talk to them *nor* they to you**. Neither can you talk to either the lifeboat or helicopter racing to your aid.

Can you tell the Coastguard **exactly** where you are? If not, neither the Coastguard nor the lifeboat or helicopter can take direction-finding bearings on your signal as they can on Channel 16.

The range to the shore of a hand-held mobile 'phone with its 0.3W output is much less than a 25W marine VHF with a much larger, mast-head aerial. The new, digital GSM 'phones are even worse than the old analogue 'phones in this respect.

You will also miss out on all the local chatter about the movement of merchant ships and you will not be able to hear navigation warnings, weather forecasts, gale warnings or SAR messages.

Lastly, can you run the mobile 'phone from the engine starter battery when its internal battery fails?

Navico's family of waterproof handheld VHF sets

The VHF GMDSS HANDBOOK

Important procedural words (prowords)	
MAYDAY	Distress call on your own behalf (Q)
MAYDAY RELAY	Distress call on behalf of someone else (Q)
PAN-PAN	Indicates urgent call concerning the safety of a ship *or person* (e.g. man overboard) (Q)
PAN-PAN MEDICO	Precedes a call to a *Coast Radio Station* requesting urgent medical advice. You will then be given a free telephone call to a doctor in the Accident Department of a local hospital (Q) NOTE: French doctors speak only French! Most others speak English
SEELONCE MAYDAY	Radio silence on Ch. 16 is imposed by the station controlling distress communications using this phrase. Silence should be automatically observed during a distress situation (Q)
SEELONCE DISTRESS	Radio silence on Ch. 16 imposed by any station other than the one controlling distress communications (Q)
PRUDONCE	A concession, at the discretion of the controlling station, to allow essential signals on Ch. 16 – even though a distress situation may not be fully concluded (Q)
SEELONCE FEENEE	End of radio silence (Q) Given by the controlling station.
SÉCURITÉ (SAY-CURE-EE-TAY)	Safety signal to indicate that a message of navigational importance is about to be sent (Q)
SAY AGAIN	Repetition required (Q)
ALL AFTER	Everything following word or phrase indicated (Q)
ALL BEFORE	Everything prior to word or phrase indicated
ALL BETWEEN... AND...	Everything between words or phrases indicated
WORD AFTER	As above
WORD BEFORE	As above
WORD BETWEEN	As above
STATION CALLING	(your ship's name or call-sign) Form of address to station which has called you but whose identification is in doubt (see below). If, on the other hand, you *think* someone is calling you, *do nothing*, but wait for the other station to repeat the call (Q)
OVER	Invitation to other person to transmit (Q)
OUT	End of conversation (Q) NOTE: As the words OVER and OUT are contradictory it is *not* correct to end a transmission with OVER–AND–OUT! (Q)
CORRECTION	The last word or phrase was wrong. This should be followed by I SAY AGAIN...
READ BACK	Repeat the message you have just received for confirmation that it was received correctly
RADIO CHECK	Tell me the strength and quality of my signal
I SPELL	I am about to spell the word just said in the International Phonetic Alphabet

Glossary

ALRS   The *Admiralty List of Radio Signals*, the primary source of all information regarding radio stations, frequencies, etc. Available from Admiralty chart agents.

asl   Above sea level.

Authority to Operate on British Ships   Normally granted when the radio operator passes the examination for the Certificate of Competence in Radiotelephony.

Call-sign   Sequence of letters and/or figures allocated to a vessel equipped with a marine radiotelephone, e.g., GABC. Always spelt out phonetically over the air, it is less prone to misinterpretation than the vessel's name.

Capture effect   On VHF, the radio locks on to the strongest signal and reproduces that to the exclusion of all others.

CB   Citizens' Band radio; the unqualified 'free–for–all' band.

Certificate of Competence in Radiotelphony   Granted to operators who have passed the appropriate examination. This is mandatory for all operators of marine radiotelephones.

Channel 16   The Calling and Distress channel.

Channel M   A private channel used by British marinas, yacht clubs, and British yachts which wish to communicate with them.

Coast Radio Station   Shore-based 'telephone exchange' which links radio-equipped vessels with the international telephone system.

C.R.0.S.S.M.A.   French coastguard service for the English Channel.

DSC   Digital Selective Calling. The new technology for calling and Distress Alerts.

Dual Watch   Electronic device fitted to marine VHF sets allowing two channels (Ch. 16 plus a selected channel) to be monitored at once.

Duplex   Dual-frequency system allowing simultaneous two-way conversation by radio.

FM   Frequency Modulation. The system by which signals are transmitted on the marine VHF band.

GMDSS   Global Maritime Distress and Safety System. The new automated Distress and calling system starting 1st February, 1999.

MHz   Megahertz, a measure of radio wave frequency in millions of cycles per second.

Private channel   Channel allocated to a particular user and therefore not available for general use.

PTT switch   Press-to-talk switch. Employed on Simplex equipment to switch from receive to transmit mode.

Public correspondence channel   Dual-frequency international channel employed for telephone system link-up designed for Duplex operation but also available for Simplex-equipped stations.

Radio station   Any radio-equipped building, ship or aircraft which has been alocated a callsign.

R/T   Radiotelephone/radiotelephony.

Selcall   Selective Calling system: a transmitted code which alerts a particular radio station.

Semi-Duplex   Used in ship/shore correspondence when the Simplex equipment of the vessel makes Simplex procedure essential despite 'Duplex' equipment of telephone user.

Ship's Radio Licence   Obligatory for all vessels equipped with marine R/T.

Simplex   Alternate transmission and reception over one or two frequencies.

Squelch   Circuit to suppress background noise.

Traffic List   List broadcast by Coast Radio Station to inform vessel that a correspondent wishes to make contact via the telephone system.

VHF   Very High Frequency; the waveband used by short-range radiotelephones.

Working channel   The channel on which business is transacted following contact on Channel 16.

YTD   Yacht Telephone Debit; the system by which the cost of a ship/shore telephone call is debited to the radio operator's own home telephone account.

The VHF GMDSS HANDBOOK

Useful addresses

Maritime and Coastguard Agency (MCA)
Spring Place
105 Commercial Road
Southampton
SO15 1EG
Tel: 01703–329100
Fax: 01703–329204

National Authority for Marine Radio Operator Certificates

Radiocommunications Agency
(Temporary address)
New King's Beam House
22 Upper Ground
London SE1 9SA
Tel: 0171–211–0212
Fax: 0171–211–0507

UK Radio Regulatory Authority

Wray Castle
Ship Radio Licensing
PO Box 5,
Ambleside
LA22 0BF
Tel: 01539–434662
Fax: 01539–434663

Issues Ship's Radio Licences

Royal Yachting Association
RYA House
Romsey Road
Eastleigh
SO50 9YA
Tel: 01703–627433
Fax: 01703–629924

Examining Body for VHF and Short Range Certificates

AMERC NAC (GMDSS)
PO Box 4
Ambleside
LA22 0BE
Tel: 01539–432255
Fax: 01539–434663

Examining Body for Restricted Operator Certificates

Radio School Ltd
33 Island Close
Hayling Island
Hampshire, PO11 0NJ
Tel: 01705–466450
Fax: 01705–466450

Training for VHF, SRC and ROC

British Telecom
Aeronautical & Maritime Billing,
PP300 26-28
Glasshouse Yard
London EC1A 4JY
Tel: 0171–843–7383
Fax: 0171–843–7362

International Accounting Authority GB14

CI Maritime Services
PO Box 601
St Peter Port
Guernsey, CI
Tel: 01481–54506
Fax: 01481–54506

International Accounting Authority GB19

Bibliography

Admiralty List of Radio Signals,
The Hydrographic Office, Taunton

Handbook for Marine Radio Communications,
Lloyds of London Press, Colchester

GMDSS Handbook,
International Maritime Organisation, London

Marine SSB Operation (GMDSS Edition),
Michael Gale, Fernhurst Books, Arundel

GMDSS for Small Craft,
Alan Clemmetsen, Fernhurst Books, Arundel

For a free full-colour brochure listing all our other books,
please write, phone or fax us at:

**Fernhurst Books,
Duke's Path, High Street,
Arundel, West Sussex, BN18 9AJ**

Tel: 01903 882277   Fax: 01903 882715